SpringerBriefs in Bioengineering

SpringerBriefs present concise summaries of cutting-edge research and practical applications across a wide spectrum of fields. Featuring compact volumes of 50 to 125 pages, the series covers a range of content from professional to academic. Typical topics might include: A timely report of state-of-the art analytical techniques, a bridge between new research results, as published in journal articles, and a contextual literature review, a snapshot of a hot or emerging topic, an in-depth case study, a presentation of core concepts that students must understand in order to make independent contributions.

More information about this series at http://www.springer.com/series/10280

Krish W. Ramadurai • Sujata K. Bhatia

Reimagining Innovation in Humanitarian Medicine

Engineering Care to Improve Health and Welfare

 Springer

Krish W. Ramadurai
Massachusetts Institute of Technology
Cambridge, MA, USA

Sujata K. Bhatia
Chemical & Biomolecular Engineering
University of Delaware
Newark, DE, USA

ISSN 2193-097X ISSN 2193-0988 (electronic)
SpringerBriefs in Bioengineering
ISBN 978-3-030-03284-5 ISBN 978-3-030-03285-2 (eBook)
https://doi.org/10.1007/978-3-030-03285-2

Library of Congress Control Number: 2018961402

This Springer imprint is published by the registered company Springer Nature Switzerland AG
The registered company address is: Gewerbestrasse 11, 6330 Cham, Switzerland

To the victims of humanitarian crises and conflict…you are not alone.
And to my parents and family, here's to making the world a better place.
– Kris
For Calestous—our dear friend, mentor, and colleague.
– Kris and Sujata

Preface

Take a look at our world today and what do you see? We could indeed answer this simple question by saying that we see a global population of 7.2 billion people, with each individual functioning as a contributing entity in the workings of our greater societies and civilizations. While indeed this macroscopic view depicts an orderly world, what happens when we look closer? When we peer into the inner workings and dynamics of our human population, we begin to see the effectors of inequalities and disparities. This is very much true when it comes to the relative health and wealth of various segments of our human population. Nearly half of the human population (3.6 billion) lives in objective poverty, living off of $2.50 USD a day, and more than 1.3 billion people live in extreme poverty—less than $1.25 USD a day. The notion of poverty is not just quantitative in nature but reflective of qualitative social components. Specifically, impoverished individuals are extremely susceptible and vulnerable to crises and conflict. Developing countries around the world that often harbor impoverished populations are generally host to weak supporting political, economic, social, and healthcare infrastructure and institutions. The fragility complex surrounding these weak supporting institutional elements in developing countries is why so many countries are so susceptible to catastrophic humanitarian emergencies and crises. But what do we mean by humanitarian emergencies and crises? A humanitarian emergency is an event or series of events that represents a critical threat to the health, safety, security, or well-being of a community or population. Our world has been host to a myriad of humanitarian emergencies ranging from the Somali famine in 1992 and Rwandan genocide in 1994 to the presently ongoing civil wars in Yemen and Syria.

From a global humanitarian perspective, more 125 million people around the world are in need of immediate humanitarian assistance and aid. While war, famine, displacement, political conflict/discourse, and natural disasters are most certainly detrimental to any population, these humanitarian crises are particularly devastating to the 50% of the global population that lives in poverty, the reason being that this population typically lacks access to fundamental resources such as food, shelter, and healthcare. This is further amplified in developing countries which often harbor highly volatile and fragile governments as well as weak economic and healthcare systems. This culmination of weak supporting infrastructure and governing bodies coupled with subsistence living amplifies the gradient of any humanitarian crisis and emergency. But what is the fundamental defining element of the human condition in these crises? That would be the preservation of human health and dignity via humanitarian medicine and innovation. The goal of any humanitarian operation is to seek the preservation of human health by providing in-the-field care and support services via humanitarian aid workers and practitioners in order to save lives. Humanitarian medicine is made up of a wide range of practices including battlefield medicine and surgery, emergency teams in disaster situations, vaccination campaigns, and public health education. Humanitarian medicine is a core component of the humanitarian emergency relief and delivery paradigm, and it is important to note that the degree of humanitarian emergencies is not limited to a short-term timeframe. Oftentimes aid delivery is viewed as a short-term initiative, whereby the delivery of aid and assistance is provided with only immediate benefits and remedies in mind. Everyone seeks a "success story." However, the fact is that the true depth of a humanitarian crisis sinks far deeper than many care to recognize, and the effects typically perpetuate and permeate society long into the future.

So how do we begin to remedy this? In this book, we focus on humanitarian innovation and fostering creative problem-solving to address humanity's most pressing problems, specifically, the ability to redefine how practitioners and crisis-affected communities themselves can pool their intellectual capital together to enhance human health outcomes in the field. This includes enhancing the interventional capacities of humanitarian medical practitioners, aid workers, and the greater communities they serve from the ground up. We specifically seek to explore the intersection of various innovation paradigms—i.e., frugal, open, and reverse innovation—in fostering novel technologies and strategies that can serve to enhance healthcare support and delivery in crises. By redefining the humanitarian medicine paradigm through the application of feasible, utilitarian technologies, we can ultimately improve and save the lives of millions more people around the world. For it is in creating solutions to help our fellow man that the purest essence of innovation is derived.

Cambridge, MA, USA   Krish W. Ramadurai
Newark, DE, USA   Sujata K. Bhatia

Abstract

Throughout our history, humanity has been plagued by a myriad of humanitarian crises that have opened the door to perpetual human suffering. This holds true in the present day, where approximately 125,000,000 people require humanitarian assistance as the result of famine, war, geopolitical conflict, and natural disasters. A core component related to human suffering experienced is that of afflictions related to human health. Each of these situations creates an impetus for morbidities and comorbidities that must be treated medically. Perhaps one of the most startling elements is that oftentimes life is lost to preventable medical conditions that were not properly treated or even diagnosed in the field. This is often due to the limited interventional capacity that medical teams and humanitarian practitioners have in these scenarios. These individuals are often hindered by medical equipment deficiencies or devices not meant to function in austere conditions. The essence of humanitarian medicine is to alleviate suffering, but the key to this strategy is to enhance the interventional capacities of humanitarian practitioners, particularly in the realm of healthcare delivery. The development of highly versatile, feasible, and cost-effective medical devices and technologies that can be deployed in the field is key to enhancing medical care in unconventional settings. But where do we begin? In this book we examine the nature of the creative problem-solving paradigm and dissect the intersection of frugal, disruptive, open, and reverse innovation processes in solving humanity's most pressing problems. We define the relative capacities of innovation processes in serving as an impetus for the engineering and development of novel technologies in redefining humanitarian medicine and aid relief. Specifically, we delve into the feasible deployment of these devices and technologies in unconventional environments not only by humanitarian agencies but also by crisis-affected communities themselves. We explore how to harness the power of various innovation processes to empower humanitarian practitioners in crisis situations as well as the very people and communities they serve. We explore this as well as the application of other innovations and technologies across multiple areas to radically improve humanitarian aid and disaster relief. In this book we take the complex challenge of developing innovative solutions for the delivery of humanitarian aid and medicine head on.

Contents

Chapter 1
The Humanitarian Relief Paradigm

When we close our eyes and picture a humanitarian crisis, we are likely to see an aurora of chaos, conflict, anger, fear, and mortality that lingers over a large group of people. However, we are also likely to see a slew of responding multilateral relief agencies and nongovernmental organizations (NGOs) that deploy and respond to catastrophe. The deployment of these agencies, whether it be the United Nations (UN), the US Agency for International Development (USAID), the World Health Organization (WHO), etc., generally follows a very long and complex interventional deployment strategy that involves numerous operating components. The aid delivery complex is generally not streamlined and often deals with a wealth of bureaucracy and dissonance between intervening agencies and the host country they are operating in. We can see how complex this process is from previous historical events such as the Haitian Earthquake in 2010 or the Rwandan genocide in 1994, where the very same agencies—i.e., the UN in these cases—that were supposed to help, actually become implicated in not providing adequate relief services. The interventional capacity of any agency is vital in not only stabilizing and promoting conflict resolution but also in the delivery of human health services via humanitarian medicine. Aid and relief practitioners and workers rely on access to equipment, services, and tools in order to provide adequate treatment and palliative care. What we will seek further on in this work is that oftentimes these individuals do not have sufficient access to the resources and materials they need to deliver care and health services to not only treat acute maladies but also chronic ones as well. But first we begin by defining the humanitarian aid complex and its deployment in real-world settings, emergencies, and crises around the world.

K. W. Ramadurai, S. K. Bhatia, *Reimagining Innovation in Humanitarian Medicine*, SpringerBriefs in Bioengineering, https://doi.org/10.1007/978-3-030-03285-2_1

1.1   Current Humanitarian Crises: Defining the Humanitarian Aid Complex

In taking a retrospective approach, we can see a stratified chronology of humanitarian crises that have afflicted distinct subsets of the global population around the world. These crises vary in scope ranging from the exodus of Rohingya Muslims in Myanmar to India to famine in South Sudan to the Haitian earthquake of 2010. One common threat among these incidences is that of human suffering, more specifically that of suffering derived not only from conflict but limited access to basic and fundamental resources food security, shelter, nutrition, protection, and health. So how do we react? This is where we first begin to define the humanitarian aid complex, which is a highly functional and dynamic entity. The goal of any aid complex is to remedy human suffering and restore access to fundamental and basic necessities that any human being in the modern era would need to survive. The purest essence of relief work is grounded in the notion of providing aid to those that need it most to enhance their quality of life in given situation or scenario. This notion, however, can become diluted by an array of multifaceted, confounding variables that challenge the principles of this basic notion. Perhaps one of the challenging elements of any humanitarian mission is the ability to deliver aid in unconventional situations that can be highly volatile and unstable. There are three key notions to any humanitarian mission, this being: scope, application, and intervention (Levy and Sidel 2008). Identifying the scope of the operation and scaling relief efforts to account for the relative number of individuals that need aid is the first step, which sets the tone for the rest of the operation. The next step involves identifying the application of the operation, how will we best delegate resources in the form of food security, protection, shelter, and healthcare to our target populous. The final step is the actual delivery of the intervention in the field and on the ground via the coordination of entities such as multilateral organizations and nongovernmental agencies.

Although we have outlined three simple steps to a humanitarian mission, what we tend to see is a complex, and often-inefficient logistical approach to these missions. This ultimately impacts not only the logistics of operations but human health. The failure to sufficiently and succinctly deploy resources can ultimately lead to a failure to deliver adequate medical care, which can ultimately impact human health. Time and time again, relief agencies are often plagued by too many moving parts and in many cases, fail to deliver direct, point-to-point care in the field. This is reflected in Figs. 1.1 and 1.2, which display huge disparities between the demonstrated need and the actual targeted delivery of aid in countries including Libya and Yemen. But what is the result of this massive dissonance? Many times, it is the loss of human life, which in some cases could be prevented with the deployment of field-ready equipment that can enhance the interventional capacity of humanitarian practitioners. This is of particular importance when it comes to health, as it is a basic tenet of the human condition and reserved right for all citizens of our world. Now imagine the health effects that these pertinent access constraints have on these individuals. Without access to food, shelter, and protection, these populations are

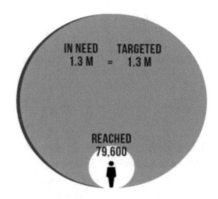

Fig. 1.1 Access constraints to humanitarian aid in Libya, where only 6% of the population has received aid. (Crisis Overview 2016)

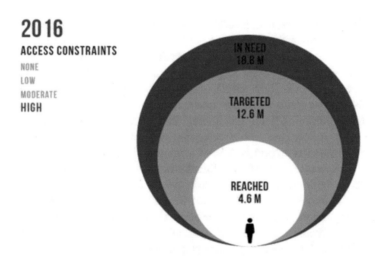

Fig. 1.2 Access constraints to humanitarian aid in Yemen, where less than a quarter of the population has received aid. (Banatvala and Zwi 2000)

exposed to medical ailments and conditions that have devastating consequences on quality of life. Incidences of pathogen infections such as malaria, cholera, and E. coli become rampant. Furthermore, simple conditions and traumas such as an injury, cut, common cold, or flu become chronic conditions that can lead to infection and even become deadly. The rapid transformation of such simple pathologies to a chronic disease states is particularly devastating in children, as they lack resistance to infection, have immature immune systems, and malnutrition can promote immunosuppression (Banatvala and Zwi 2000).

Table 1.1 Types of humanitarian emergencies. (Rodrigues Santos 2015)

Type of disaster	Natural	Technological	Complex
Sudden-onset	Tropical storm Earthquake and Tsunami Landslide	Industrial accident Road/air-traffic Accidents	Terrorist attack Conflict Epidemics
Slow-onset	Drought Famine	Air and water pollution	Political or refugee Unrest Conflict Poverty

In defining the humanitarian aid complex holistically, identification of the common sources of humanitarian emergencies is imperative. In Table 1.1, there are two types of disasters, sudden-onset and slow-onset, which are characterized by how quickly the disaster emerges and afflicts a population. Within these two types of disaster, we can further classify the realm in which it occupies, specifically that of being natural, technological, or complex in nature. Perhaps one of the most intriguing elements related to this classification of disasters is that they often are intertwined, in which elements of a sudden-onset disaster such as a natural disaster can foster and perpetuate the development of slow-onset conditions such as famine, conflict, and poverty. Each one of these classifications calls for a distinct response, yet inevitably, these responses are geared toward protecting the population and preserving the health of individuals. Oftentimes this simple goal becomes difficult to sustain as many operations provide short-term immediate care, but leave individuals vulnerable to returning to an environment that is unstable. Immediate care is certainly a staple of any operation as mitigating the direct threat to human health and safety is of the utmost importance. Yet this mind-set is flawed in that many countries afflicted with disasters are often politically unstable and prone to conflict, further perpetuating these crises. For example, poverty, famine, and drought are entities that cannot simply be remedied immediately via a short-term intervention. These situations require information dissemination, technological innovation, and human capital development in order to create a sustainability complex that can alleviate the impact of future disasters and enhance preparedness.

There are huge disparities in access and delivery of humanitarian aid, particularly in areas of high conflict, natural disaster, and political dissonance. One of the most challenging elements to any operation is not only the ability to effectively identify and intervene in a humanitarian crisis. But how do we ensure that aid is distributed equally and effectively? The answer is that we cannot, as oftentimes in areas of high conflict and duress, the ability to rapidly disseminate aid is a priority. Even with the rapid dissemination of aid, the sustainability complex of these operations is often limited within the short run, which is why there is such a huge disparity in the number of people reached and those in need. The sheer scale and size of many humanitarian missions oftentimes mean that an aura of "controlled chaos" resides that tests the physical and mental capacities of many humanitarian actors.

While there are indeed certain elements that are out our control in delivering aid, many multilateral organizations such as the World Health Organization (WHO) have created innovative logistical systems to lift the burden on field operations and effectively organize agencies so that they can maximize their respective operations and tasks in the field. Such approaches include the "cluster approach" enacted in 2005, as a major reform of humanitarian coordination (Rodrigues Santos 2015; What is the Cluster Approach? 2017). The Humanitarian Reform Agenda (HRA) introduced this approach in order to enhance predictability, accountability, and partnership among various humanitarian agencies. The cluster approach involves organization groups of humanitarian organizations, both UN and non-UN, into "clusters" in each of the main sectors of humanitarian action, such as water, health, and logistics (Fig. 1.3). They are designated by the Inter-Agency Standing Committee (IASC) and have clear responsibilities for coordination and delivery of services (Crisis Overview 2016).

With logistical platforms such as the cluster approach, interagency cooperation is maximized in each of the 11 sectors of humanitarian action which include camp coordination, early recovery, education, emergency telecommunication, food secu-

Fig. 1.3 The Cluster Humanitarian Approach. (What is the Cluster Approach? 2017)

rity, logistics, nutrition, protection, shelter, water and sanitation, and health. While we could indeed explain the relative interventional capacity of each of these sectors, for the sake of this book, we focus on one particular sector, this being health. Health is a fundamental element of the human condition and serves as a metric for how well a humanitarian mission is executed and delivered. Health is a dynamic entity in crisis situations as it is influenced by each of the other sectors that we previously outlined. The health status of a population is innately tied to its social and economic prosperity, security, resource accessibility, and nutrition. Protecting health serves as a sort of metaphysical armor for protecting the future of humanity and generations to come. Given this notion, how do we deliver effective, yet efficient care in humanitarian settings? The key to this lies in not only delivery of healthcare services via aid workers but also local public health practitioners and proxies that are symbiotic with the host population and can truly discern the needs of the local populous. Even with the added benefit of engaging these practitioners, the ability to define the most critical medical elements to be deployed in humanitarian operations is vital in order to properly meet the needs of conflict and disaster victims.

1.2   Medical Treatment in Unconventional Settings: Meeting the Needs of Conflict Victims

Perhaps one of the largest impacts of war, natural disasters, or any situation requiring humanitarian assistance is on human health. Refugees and internally displaced people typically experience high mortality immediately after being displaced (Levy and Sidel 2008). The most common causes of death include diarrheal diseases in the form of cholera and dysentery, measles, acute respiratory infections, and malaria, which are often exacerbated by malnutrition (What is the Cluster Approach? 2017). Communicable diseases are highly transmittable in emergency and relief scenarios due to mass population movement, overcrowding, impoverishment, poor sanitation and nutrition, as well as limited access to healthcare (Connolly et al. 2004). Furthermore, acute disabilities related to injury are likely to perpetuate and become chronic in nature, requiring long-term healthcare. These conditions are most detrimental when they afflict the most vulnerable of individuals including the elderly and of particular significance, children. Of the ten countries with the worst mortality rates for children aged under 5 years, seven are affected by complex emergencies (Black et al. 2003). Upon acknowledgement of this, the goal of this book is to identify novel strategies, technologies, and innovations that seek to sufficiently address the needs of conflict victims and preserve health and quality of life.

More than 200 million people around the world live in countries where complex emergencies affect not only refugees and internally displaced people but the entire population. This is primarily due to the fact that these countries have healthcare systems and infrastructure that are compromised due to a loss of health staff, damage to infrastructure, insecurity, and poor coordination (Levy and Sidel 2008).

These weak and fragile healthcare systems are extremely vulnerable to catastrophic collapse in complex emergencies, which effectively hampers prevention and control programs. This causes a consequent rise in communicable diseases such as tuberculosis, cholera, dysentery, as well as vector-borne diseases such as malaria, trypanosomiasis, and yellow fever, and even vaccine-preventable diseases such as measles and pertussis (Connolly et al. 2004; Levy and Sidel 2008). These factors are further compounded by unstable and weakened governments—sometimes even anti-government forces—and UN agencies such as the WHO and UNICEF as well as nongovernmental organizations for delivery of health services (Levy and Sidel 2008). The provision of clean water, food, shelter, and security becomes elements of the utmost importance, as they function as direct determinants of the relative health outcomes of disaster victims. Without the provision of these basic elements, individuals become exponentially more susceptible to dying from treatable and preventable maladies. These elements further prevent the cyclical nature of infection and illness, whereas individuals of whom are treated return to environments that may not have adequate food, clean water, shelter, and security, which makes them susceptible to becoming ill again in a perpetual fashion. What we see here is that health is innately tied to all conditions in humanitarian settings and is perhaps the most fundamental element that governs the outcomes of relief operations.

In developing interventional strategies for providing targeting and effective medical treatment in unconventional settings, perhaps the most critical notion is that of large-scale action in delivering care during the acute phase. The highest morbidity and mortality often occurs during the acute or initial phase of a conflict emergency, in which death rates of over 60-fold the baseline have been recorded in refugees and displaced people, with over three-quarters of these deaths caused by communicable diseases (Connolly et al. 2004; Levy and Sidel 2008). The deadliness of the acute phase is often attributed to the culmination of multiple elements that create the "perfect storm" for infectious disease outbreak and rapid morbidity among refugee and internally displaced individuals. Oftentimes during the acute phase, a resurgence of previously controlled diseases such as malaria and trypanosomiasis occurs (Connolly et al. 2004). Furthermore, the emergence of drug resistance driven by improper and incomplete use of drugs such as bacillary dysentery and multidrug-resistant tuberculosis occur (Connolly et al. 2004). These countries affected by conflict play host to new disease emergence because of delays in detection and characterization of new pathogens and their widespread transmission before control measures can be implemented such as in the case of monkeypox in Democratic Republic of the Congo (Rodrigues Santos 2015). Beyond infectious diseases, medical treatment in the acute phase allows for medical practitioners to prevent conditions and maladies from developing into chronic conditions.

In countries with previously weakened healthcare infrastructure, limited human capital, and poor economic development, the confluence of these infectious disease outbreaks in conjunction with acute and chronic traumas and conditions, creates a deadly outcome for the indigenous population. This further hinders the ability for adequate and sustained humanitarian operations, as rather than directly focuses on one specific disease pathology; oftentimes there is a merging of political disso-nance, social and cultural upheaval, and economic disparities, which means that operations have to tackle an entire system at play. With this in mind, how do we adequately meet the needs of conflict victims? The answer to this seemingly simple question is complex, since current operations oftentimes fail to meet even the short-term needs of conflict victims, so how could we even focus on the long-term needs? Furthermore, in looking closer at the disease pathologies and health conditions of the poor, these are often conditions classified as "orphan diseases," which are com-mon disease that are generally ignored (such as tuberculosis, cholera, and typhoid) because it is far more prevalent in developing countries than in developed countries (Aronson 2006). This means that these conditions are harder to treat in conflict situ-ations, as aid agencies must rely on a steady stream of vaccines to be delivered from developed nations, where they are researched and developed. This situation coupled with limited infrastructure and human capital development means that many con-flict areas, which occur in developing countries, have an effective crutch and reli-ance on either foreign aid agencies or developed nations for medical treatment.

What we begin to see here is the need for more "in-field" and "on the ground" solutions for effective aid delivery and distribution of medical care in these volatile and unconventional relief settings. An impetus for this notion would be that of for-eign aid agencies, which would not only treat the conditions or maladies that they were sent to alleviate but also deploy human capital development and medical tech-nologies, suited for development in various settings. By this we mean the deploy-ment of frugal innovation and basic, engineered medical devices that can not only effectively diagnose and treat conditions but serve as a platform for further develop-ment and intellectual inquiry. We seek not only to enhance the interventional capac-ity of humanitarian practitioners but also that of the native population in the effective treatment and care of medical conditions in the field. This approach is unique in that it expands the realm of humanitarian medicine to include all individuals as stakeholders.

Perhaps the most important stakeholder that is almost entirely overlooked in many operations is that of the community health worker and community-based health programs. These individuals and greater programs serve as a vital conduit between humanitarian operators and the local population that is being served and are critical in extending healthcare delivery and improving health outcomes (McCord et al. 2013). Evidence has supported that strong community-based health programs can reduce infant and child mortality and morbidity and provide low-cost interventions for common maternal and pediatric health problems while improving the continuum of care (McCord et al. 2013). These community-level programs are extremely effective in addressing the most common causes of pediatric mortality and morbidity, such as pneumonia, diarrhea, and malnutrition (Lewin et al. 2010). These individuals in these programs are well-acquainted with the social, cultural, and geopolitical components of their respective populations and can serve as a vital source of knowledge. Perhaps one of the biggest weaknesses of foreign aid agencies is the formal understanding of the general population that they are helping. One of the most prominent examples of this is that of the Ebola outbreak in West Africa.

Upon delivery of aid, many medical and aid workers were unaware of the ritual burial practices of the communities throughout West Africa. These burial practices were largely responsible for the rapid dissemination and outbreak of Ebola, in which entire families would come in contact with the deceased individual and become infected (Alexander et al. 2015). Community health workers were vital in relaying these practices to operators in the field, which prevented the further propagation of the virus. Properly outfitting community health workers and humanitarian practitioners in crisis situations is paramount. With sufficient access to the novel technologies and innovations that we later explore in this book, we can redefine the humanitarian medicine treatment paradigm.

In further discussing our ability to meet the needs of conflict victims, a fundamental understanding that health is not an entity of short-term focus, but rather a long-term functioning agent, is pertinent to revising interventional strategies. Perhaps one of the most overlooked concepts is the transition of medical conditions and ailments from acute to chronic conditions. More specifically, the development of chronic illness as what we term as a "perpetuity" or a perpetual state of illness or affliction. This is often forgotten as the functional capacity of many humanitarian aid missions is to provide care and delivery of aid within a small window of time, with little regard to the state of affairs that ensues after an operation is completed. While it is certainly true that the essence of humanitarian aid is to not rewrite policy or change the political state of affairs of any country, however, when it comes to health, we argue that it is indeed the responsibility of responding organizations to leave the countries that they serve in a state than when they first arrived. Improving the health of a populous afflicted by conflict can serve as an empowering agent of change, as displaced individuals are perhaps the most vulnerable group. A healthier population, especially in a state of displacement, can improve morale, enhance survival, and restore a state of balance rather than that of despair.

1.2.1   Health Is Wealth: Avoiding Chronic Illness as a Perpetuity

The acute to chronic condition translational gradient could be the most critical concept in delivering palliative care in any setting, but is even more important in conflict emergency situations, that often occur in developing countries. Individuals in these countries oftentimes work in agricultural sectors and rely on physical labor as their source of income. Acute traumas have the distinct propensity to manifest themselves into severe, life-altering chronic conditions that not only can impact the relative health of an individual but also their social and economic contributions to society as a whole. For example, an agricultural laborer falls victim to natural disaster and faces an unintentional injury such as a broken arm, if the bone is not properly treated and set during the acute phase, the patient would most likely not have this malady addressed in the future due to either fiscal constraints or access to reliable healthcare. This eventually turns into a situation where the bone does not heal properly, in which the simple nature of a broken arm, becomes chronic in nature and a disability, which inhibits the patient from working and providing for their family. Individuals with physical disabilities are likely to face a multitude of barriers that will prevent them from effectively participating in their societies. Furthermore, disability and poverty are closely linked, in which preventing this transition from acute to chronic condition has social, economic, and health consequences.

We can classify injuries typically encountered in humanitarian missions as intentional or unintentional injuries. Intentional injuries refer to injuries caused by direct agents including gender-based violence, assault, self-harm, and suicide, which accompany such humanitarian situations related to war and conflict (Chan 2017). Unintentional or unplanned injuries are typically associated with natural disasters and include injuries from fires, falls, motor vehicle accidents, as well as work-related incidents. Like all conditions, untreated wounds and improper injury care can result in avoidable mortality, morbidity, and permanent disability, which can impact an individual's post-disaster livelihood (Chan 2017). Deaths represent only the proverbial "tip of the iceberg" of the true burden of unintentional injuries, as many injury events are not directly fatal, but function as indirect afflictions that eventually result in chronic conditions which affect people throughout their lives (Chandran et al. 2010). Over 90% of unintentional injury deaths occur in people in low- and middle-income countries (LMICs). When standardized per 100,000 population and compared with high-income countries, the death rate is nearly double in LMICs (65 vs. 35 per 100,000), and the rate for disability-adjusted life years (DALYs) is more than triple in LMICs (2398 vs. 774 per 100,000) (Chan 2017; Chandran et al. 2010). Thus, individuals in LMICs are more prone to death due to injuries and associated complications. With the combined impact of disproportionately early mortality, the need for critical and costly medical care, and the risk for extended periods of disability, traumatic injuries present a staggering cost burden to individuals, families, the medical system, and governments (Chandran et al. 2010). Perhaps an even more concerning element is that these health outcomes and metrics represent a pre-humanitarian emergency and crisis situation. We can clearly see the massive dissonance of healthcare services provision and global burden of disease (GBD) which afflicts LMICs disproportionately. However, in humanitarian crisis and emergency situations, we can see a complete degradation in health outcomes and exponential increasing of the vulnerability complex of already previously vulnerable populations. This means that individuals in these scenarios are for more susceptible to succumbing to various maladies, thus increasing the functional burden of disease.

In humanitarian missions, workers and practitioners not only face the treatment of acute maladies as well as the acute to chronic condition paradigm but are often exposed to medical conditions that are already chronic in nature. These typically manifest themselves in the form of noncommunicable diseases such as arthritis, diabetes, obesity, cancer, and an array of other pathologies. These conditions often take a backseat to acute treatments in the field, in which the individuals facing these complex pathologies are often left to their own devices in receiving palliative care after humanitarian relief is delivered. In light of this startling notion, why do humanitarian practitioners neglect chronic diseases in the first place? The reasons why humanitarian practitioners neglect chronic disease post-disaster include (adapted from Chan 2017):

- Diminished awareness, cooperation, coordination, and sustainability
- Lack of resources as well as standardized protocols for managing chronic diseases in relief settings
- Lack of knowledge of local demographics as well as relevant skills and expertise to detect, manage, and treat chronic diseases in post-disaster settings
- Resistance to change by local health systems and stakeholders who lack understanding of the problem

A distinct pattern indicates that population groups that display pre-existing chronic medical problems tend to be worse off in post-disaster and conflict situations. These problems are once again rooted in the fact that most post-disaster medical relief focuses primarily on the provision of acute medical treatment and the control of communicable diseases (Chan and Sondorp 2007). Large rates of mortality, destruction of health infrastructure, and political instability have often impaired the ability of both government and relief agencies to provide relief beyond essential lifesaving procedures (Chan and Sondorp 2007). Furthermore, relief workers might also unknowingly complicate the medical condition of population groups through their insensitivity toward specific needs of patients with chronic medical conditions. For instance, adverse drug interactions may go unnoticed and dietary needs of diabetes and hypertensive patients may not be addressed in food distribution in disaster relief operations (Chan and Sondorp 2007). Unless sensitivity and knowledge of relief workers toward chronic medical needs increase, medical interventions will remain less efficient and effective.

An interesting facet related to this notion is that disaster situations not only cause massive loss of human life but also provide a distinct opportunity for underserved populations to receive access to external resources that harbor improved medical services and care that may not be available in that locality (Chan and Sondorp 2007). Many relief agencies often remain at a disaster-affected region for more than 6 months, which creates the focal opportunity to not only deliver short-term care but build the foundation for the delivery of long-term palliative care (Chan and Sondorp 2007). These humanitarian relief agencies are often equipped with the human capital, technical expertise, and resources which could be directly geared toward the restoration, reconstruction, and improvement of health services in disaster- and conflict-affected areas. This could be effectively implemented via the establishment and use of low-cost interventions for disease prevention and reduction of chronic medical complications alongside emergency primary healthcare services provided during the relief phase of the operation. Furthermore, this creates a situation that allows for knowledge transfer between humanitarian practitioners—i.e., doctors, nurses, and medical support staff—and domestic community health workers that are from the crisis-afflicted area. This knowledge transfer includes the flow of medical information, techniques, as well as equipment and devices that can be vital in future treatment of patients. During the facilitation of this transfer, the concept and methods of frugal engineering and innovation are a focal element that has the power to radically improve patient outcomes and treatment. We delve further into this fascinating paradigm at the end of this chapter, but it is a concept that can certainly serve to improve short- and long-term palliative care solutions in humanitarian crises and emergencies.

1.2.2   The Humanitarian Paradox: What Happens When We Leave?

In looking at the delivery structure and capacities of humanitarian relief, the degree of socioeconomic development plays a fundamental role in the provision of medical services and health resources. But a profound question still remains, what happens when humanitarian operations cease and leave? The post-disaster situation is as important, if not, even more important that the initial phase of the operation. Specifically, it is during the post-disaster phase that we ask ourselves, did we leave a population better off than when we found it? These questions are important to ask, as future health and wealth of entire populations and countries are dependent on the wellbeing of their citizens. As previously mentioned, conflicts tend to disproportionally affect the most vulnerable, and even if disaster medical relief groups attempt to address chronic medical service during relief interventions, they would have to overcome the lack of pre-existing local service structures, including human resources and technical knowledge, in order to sustain services created for medical treatment during a relief operation. This set of constraints alone usually deters emergency relief agencies to venture into the arena of managing noncommunicable chronic conditions. For example, despite the fact that 25% of people using health facilities following the earthquake of Kashmir, Pakistan, were older people, none of their chronic health conditions were managed because these needs were not characterized or targeted during the initial relief assessment. The continued medical treatment of individuals after relief operations leave is a challenging situation that is further perpetuated by the dissonance in governments to adequately address these problems. Humanitarian missions should set the stage for an effective paradigm shift in medical treatment, so that a "crutch" is not created where developing countries rely on aid as a perpetuity from developed countries.

Conflict and disaster relief planning should work to provide an integrative approach to adequately address the health needs of patients with a variety of maladies including noncommunicable diseases during emergency. However, in LMICs, fragmented medical services and limited health resources typically limit access to appropriate care for people with both acute and chronic medical problems. This condition is further exacerbated by the experience of conflict or disaster, which can radically alter the health-delivery landscape. While it is critical to respond quickly to save lives and prevent suffering, obtaining valid information to make evidence-based, appropriate, and relevant relief decisions is just as important, especially for setting a transitionary stage for healthcare in countries (Chan and Sondorp 2007). Healthcare needs are those needs that can benefit from a service or intervention along the pathway of care, namely, health protection, health education, disease prevention, diagnosis, treatment, rehabilitation, and terminal care. A health needs assessment is an effective tool in identifying the unmet healthcare needs of a population and making changes to meet these unmet needs. The problem is that traditional needs assessments tend to focus on identifying health risks rather than health needs (Chan and Sondorp 2007). These health "needs" assessments focus on minimizing potential health risks or hazards (such as possible disease outbreaks) instead of supporting ongoing chronic medical/health needs which have been present prior to the conflict or disaster as well as those that will perpetuate afterward.

It is also important to note the indirect factors associated with humanitarian relief and that the factors of health are indeed multidimensional. Ill health can be caused by a myriad of agents including the lack of basic necessities for healthy living such as access to clean water and sanitation or food and nutrition, poor environmental factors, as well as inadequate housing and security (Chan and Sondorp 2007). Relief operation recommendations drawn from single sector assessments will be inadequate to address all aspects of underlying health needs, and it is vital to identify the capacity and performance of the local health services. In humanitarian crises, competition over limited resources, the lack of know-how, and the absence of institutional strategies to deal with chronic medical problems provide very poor incentives for medical care workers to deal treat non-acute and chronic conditions that will continue beyond the operation. In addition, in developing countries where pre-disaster health services are limited, it will be unlikely to find relevant local staff with adequate knowledge of drug use and skills to support the management of chronic disease. The lack of appropriate human capital, supporting structures, history of multidisciplinary collaboration, and government's policy that would be required to sustain services beyond the emergency will often hamper the management or development of relevant chronic disease strategies post-disaster.

Furthermore, the absence of relevant field-friendly clinical diagnostic criteria, proper medical equipment, and devices also pose additional challenges to provide not only acute care but also chronic medical service in resource-deficit settings. Fundamentally, many disaster medical interventions are acute in nature and targeted on short-term outcomes. Unless relief agencies or governments have strong underlying development ethos or policies that encourage community participation in the process, the concept of planning the long-term strategies and sustainable exit planning are simply neglected. Table 1.2 displays the factors that impede proper medical care in post-disaster and conflict situations.

In going back to our original question at hand, how do we best meet the needs of conflict victims? The answer to this is that we enhance the interventional capacity of operators in the humanitarian field and acute care settings, but what does this entail? When dealing with unconventional situations such as that of humanitarian crises, the ability to properly retrofit healthcare operators in the field is paramount to delivering acute and chronic as well as lifesaving care. Specifically, we mean the

Table 1.2 Factors that impede medical care for chronic conditions during the post-disaster and conflict medical interventions

Phase	Factor
Pre-disaster	No services available
	No human or technical resources
	Chronic medical needs not included into disaster preparedness
	No government policy priority
Impact	Acute needs take priority
	Relief worker unawareness
Needs assessment	Risk-based vs. need-based
	Acute need focus
	Lack of sensitivity of relief workers toward chronic care
Rehabilitation	Opportunity cost in investing limited resources in chronic medical services
	Absence of institutional strategy to deal with longer-term strategies
	Absence of clinical diagnostic criteria in the field and availability of medical device resources

Adapted from Chan and Sondorp (2007)

deployment of low-cost, highly efficient, and effective frugally engineered medical devices in conjunction with novel strategic initiatives to treat the most basic and pertinent acute conditions. These devices can be further used to enhance the practicing capacities of community health workers who will be in charge of medical care after humanitarian operators leave. Providing adequate medical devices in the field are a must in order to vastly improve the effective capacities of practitioners to treat and diagnose patients in the field. In order to further develop this idea as well as the devices themselves, we must first briefly paint a picture of the disparities in medical device technologies in these environments and what is needed most to enhance palliative care.

1.3   Disparities of Healthcare Services in Conflict and Disaster Areas

What we have seen so far is that deployment of humanitarian aid and relief is a complex entity that is dynamically influenced by the social, economic, and geopolitical forces. While indeed we can acknowledge the complexity of this, the purpose of this book is to reflect upon how we can deploy novel technologies and innovations to improve human health. The current delivery of healthcare in such unconventional situations must certainly be revitalized and enhanced with medical technologies and innovations that are highly versatile, affordable, and ultimately enhance the outcomes of patients treated. But before we delve into these devices and applied technologies in this book, we need to outline the disparities of medical device equipment and attainment in unconventional conflict and disaster settings. When we look at humanitarian operations in countries throughout the world, photos of large food drops, helicopters, refugee camps, UN peacekeepers, and an array of nonprofit and nongovernmental agency workers providing relief services are depicted. Many times, the public perceives the deployment of these individuals and resources as being fully autonomous and executed with precise oversight, but in many cases, this could not be further from the truth.

We can look at the Great Lakes refugee crisis in 1994 after the Rwandan genocide, where a mass exodus of over two million Rwandans into neighboring countries of the Great Lakes region of Africa occurred (Malkki 1996). Many of the refugee camps set up by the United Nations in the African countries of Zaire, Burundi, and Tanzania, ended up becoming militarized and penetrated by the interahamwe and genocidiaries who were directly attributed to committing crimes against humanity. Within these camps, which were supposed to serve as safe havens, thousands more were killed and subjected to acts of terror. Medical relief services were simply not adequately equipped nor readily able to face the exponentially growing scale of medical emergencies including prominent disease pathologies such as bloody and nonbloody diarrhea, sexually transmitted diseases (STDs), malaria, cholera, measles, meningitis, and acute respiratory infections, and HIV/AIDS (Connolly et al. 2004). Thousands of people died from easily treatable conditions, but the ability and capacity of medical and aid workers to deliver the basic elements of palliative due to a lack of personnel, situational instability, as well as lack of access to medical devices such as diagnostic equipment and basic life support devices. We can see in these complex and highly volatile situations that an integrative approach must be garnered in which surveillance and security in conjunction with proper outfitting and availability of medical personnel and equipment is critical in enhancing the treatment efficacy of conflict victims and refugees. It is startling to see that this situation can indeed be mirrored in a host of other humanitarian emergencies and crises including the Ebola outbreak in West Africa, the war in Syria, as well as conflict and civil war in Mali, Sudan, Yemen, and Libya.

A common thread among many humanitarian crises that of the presence of extremely weak health infrastructure which harbors significant gaps in resource allocation and staffing, ultimately impeding adequate healthcare service access and delivery. These systems are extremely vulnerable before a crisis; thus during and after a crisis, they become catastrophically deteriorated and, in some cases, completely decimated. Even if excellent care is rendered by humanitarian operators in the field, when they leave, the morbidities and comorbidities faced by the populous are likely to become exacerbated, chronic, and lead to poor long-term health outcomes including death, as the supporting healthcare infrastructure cannot handle these needs. A prominent example is depicted by the African country of Mali, in which the destruction and looting of healthcare facilities in northern Mali—an area controlled by anti-government factions—has significantly reduced access to basic health services. Approximately 90% of community health centers in the regions of Kidal, Gao, and Timbuktu are not functional due to the departure of health personnel who have fled the violence as well as a majority of medicines and other medical supplies being looted or damaged (Bleck and Michelitch 2015). Healthcare facilities in conflict areas such as Bamako, Mopti, and Gao regions have reported hundreds of cases of traumatic injuries and mortalities, in which this is further exacerbated by the risk of disease outbreaks due to the massive population migration in conjunction with precarious living conditions (Bleck and Michelitch 2015). What we see here is a frail regional healthcare landscape that simply cannot adapt and react to discourse in the form of domestic terrorism and civil discourse. While this is certainly true in any context regardless of country, what we see is a capacity failure in the ability to treat patients due to a lack of three fundamental elements: human capital, resources, and infrastructure, human capital being medical workers, resources being the healthcare equipment, and infrastructure being the hospitals and clinics. This systematic frailness of the healthcare infrastructure ultimately gives way to a tangential shift in not only the ability to treat acute maladies in the form of injury and traumas of a small group of people but also garners further susceptibility in the ability to treat massive disease outbreaks that often occur with large population migration/ fluxes. But why is it that conflict victims and refugees are so susceptible to poor health outcomes? In order to characterize the deficiencies in healthcare, we must first fully characterize why refugee and conflict victims are so susceptible to illness and disease. There are several common causes of morbidity in crisis situations which include (adapted from Refugee Health 1995; Toole and Waldman 1993):

- Overcrowded living conditions: facilitates increased infectious disease transmission
- Poor nutritional status: lack of food before, during, and after displacement, which ultimately contributes to weakened immune system
- Inadequate clean water access and provision
- Poor environmental sanitation
- Inadequate shelter

These five elements can lead to not only the development of acute illness and disease but can actually turn easily treatable conditions, such as diarrheal illness, into a chronic pathology, which can ultimately lead to poor health outcomes and even death. This means that while short-term palliative care can be rendered in these settings, ultimately the relative health outcomes are hindered by the environment that refugees and conflict victims find themselves in. In looking closer at this predicament, who is the most susceptible and vulnerable to these conditions? There is ample evidence confirming that access to effective healthcare is a major problem in the developing countries around the world. Millions of people suffer and die from conditions for which there exist effective interventions, but in conflict settings this is further amplified. Looking closer, there is one specific group that are the most vulnerable to the health consequences of conflict—children. There are three specific diseases—diarrhea, pneumonia, and malaria—which are responsible for 52% of child deaths worldwide (O'Donnell 2007). Specifically, children under 5 years of age constitute approximately 20% of a refugee population and are the group at greatest risk (Refugee Health 1995; Toole and Waldman 1993). Among the Kurdish refugees on the Turkey-Iraq border in 1991, over 60% of all documented deaths occurred among children under 5 years of age (Refugee Health 1995). These diseases discussed are highly prevalent in conflict areas, as temporary settlements established for refugees or displaced individuals, but thousands of people in close proximity, with poor sanitation and healthcare resources. What is astonishing is that each one of these diseases has an effective prevention and effective treatment, the disparities lie in the deployment of medical treatment primarily due to lack of medical professionals, resources, and services in these conflict settings. For example, in 2009, Syria had 29,927 doctors, and in just 6 months, that number was halved due to the threat of persecution and developing civil war as shown in Fig. 1.4 (Annual Report 2009). The radical depletion of medical practitioners leads to immediate large gaps and deficits in the skills and numbers of healthcare personnel available to serve the civilian population, which is already under duress from open conflict and aerial bombings (Lipshultz 2018). However, a deeper look at this gap reveals a disparity between government-controlled areas and nongovernment-controlled areas.

In 2015, the nongovernment-controlled region of Eastern Aleppo had a doctor-to-patient ratio of 1:7000, in which only just 5 years earlier this ratio was 1:800 (Fouad et al. 2017). In this situation, there lies a massive disparity in healthcare coverage, in which 31% of Syrians live in areas with insufficient health workers and 27% live in areas with a total and complete absence of healthcare workers (Fouad et al. 2017). War-conflicted countries demand heightened attention to healthcare systems as conflict often causes a lack of healthcare workers, inventory stock-outs, and delays in treatments and vaccinations leading infectious disease outbreaks and epidemics (Fouad et al. 2017; Toole and Waldman 1993). In conflict situations, many times healthcare infrastructure and supporting entities such as hospitals and ambulances are specifically targeted, in which the burning and looting of hospitals

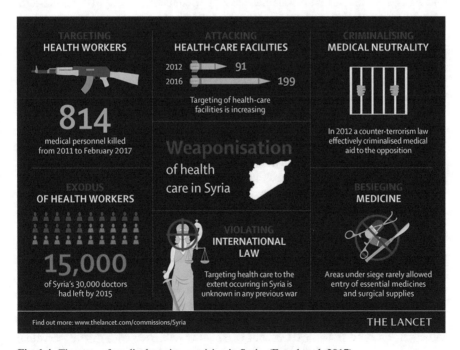

Fig. 1.4 The state of medical service provision in Syria. (Fouad et al. 2017)

and clinics, as well as attacks on medical staff and patients has occurred in facilities in countries such as the Democratic Republic of the Congo, Iraq, Afghanistan, Mali, Syria, South Sudan, Sudan, and Yemen (Lipshultz 2018). These attacks create a huge transition of the burden of care from the conflict country's healthcare system to responding humanitarian workers and teams. These aid workers are in many cases the only point of critical medical care, as all supporting structures in conflict areas are damaged or completely eradicated. The coalition also found that the passage of ambulances, medical supplies, or patients seeking care were routinely restricted in Central African Republic, Mali, the Occupied Palestinian Territory, Syria, and South Sudan (Lipshultz 2018).

The provision of immunization, nutrition programs, basic curative care, oral rehydration therapy, and family health services provides a foundation for care in conflict scenarios and emergencies. But this care is indeed not only dependent of the medical providers themselves but also the internally displaced and refugees (Refugee Health 1995). The involvement of the refugee community in its own healthcare is actually essential, not only for the effectiveness of treating maladies for the short term but also in preparing these individuals for a time when the community will eventually have to support its own primary health services (Refugee Health 1995; Toole and Waldman 1993). The deployment of refugee community health workers in conjunction with health staff and involvement of the refugee community leadership is in fact vital in enhancing aid delivery and coordination, ultimately improving care and saving lives. This can not only enhance the short-term palliative care of individuals in disaster situations but actually systematically improve longer-term care with supporting stakeholders in healthcare investment and development, prompting the equitable distribution of health services (Orach 2009). In doing this, the role of education is critical as many times conflict victims and refugees often turn to self-treatment; thus local leaders must be educated on the treatment remedies for common illnesses and the basics of infection prevention (Orach 2009). Obviously in active conflict, crises, and warfare, this is quite difficult to do; however, humanitarian agencies should feel obligated to train and supply refugee leaders to respond to emergency situations while setting up clinics and hospitals.

Many times, in humanitarian situations, there is a dynamic interplay among state-run healthcare systems and aid agencies. However, in looking at many conflict situations, what we tend to see is that existing state-run healthcare facilities are often unable to meet demand or even operate, making humanitarian aid the only way to provide care (Gamble 2018). This creates a problem of user reliance, where an entire country or region is fully reliant upon aid services—which of course is understandable for the short-term—but the problem is that aid agencies often do not stay for the duration of time it takes to fully rebuild a country's healthcare infrastructure. The means that in post-conflict situations, conflict victims, internally displaced individuals, and refugees who have come to rely on health services garnered by aid agencies, cannot access these services as their host country's health infrastructure is not fully rebuilt. This dissonance creates a highly vulnerable situation whereby individuals cannot receive the palliative care they deserve are rendered into a sort of "limbo" situation between foreign and state agencies. Furthermore, in many conflict areas and localities, the geographic, social, and economic boundaries between relief/refugee camps and host settlement have become essentially "blurred" (Leaning et al. 2011). This phenomenon has been seen in many situations including refugee settlements in Uganda, Zambia, and Yemen, and this growing phenomenon of "urban refugees" has a myriad of implications for policy, operations, and programs (Leaning et al. 2011). With regard to humanitarian information systems and operational capacity, it is difficult for agencies to keep track of people when they move to urban areas in order to adequately assure that they are receiving minimum levels of care. Furthermore, secondary and tertiary care services are more developed in urban settings; therefore, more complicated and expensive cases often present in urban refugee situations (Leaning et al. 2011). For example, chronic disease pathologies are more frequently diagnosed and treated among refugees who are located in urban dwellings compared to refugees who have fled a country and are situated in more remote areas (Leaning et al. 2011).

There is a paradigm shift occurring in medical services supplied in conflict and humanitarian crises. Increasingly, refugees and internally displaced individuals are often displaced for periods longer than a year; thus their relative health needs extend beyond the basic primary healthcare treatments provided by humanitarian agencies. As agencies have become more successful in providing basic healthcare, populations—specifically refugees—have had increases in life expectancy, creating older age groups where chronic illnesses become more predominant (Leaning et al. 2011). These chronic conditions required different palliative care measures which are typically not present with acute care initiatives harbored by most aid agencies. What is important to note in these situations is that damaged public health infrastructure and poor access to care can prevent the long-term improvement of population health and influence health outcomes for future generations. In particular, in post-conflict situations, noncommunicable diseases such as diabetes, cancer, and heart disease can surge, representing a tangential shift from acute pathologies to more chronic conditions. While indeed short-term humanitarian action can help to stabilize health crises, the broader development of fragile health systems requires a more sustainable mechanism for addressing long-term population health, disorders, and diseases. In our next chapter, we explore the use, development, and deployment of novel medical technologies fostered by frugal engineering and innovation—a concept we explore next.

1.4 Humanitarian Innovation and Frugal Engineering: A Social Perspective

The term "humanitarian innovation" refers to the role of technology, processes, products, new forms of partnership, and the deployment capacities of crisis-affected people (Betts and Bloom 2014). This term is an umbrella term that encompasses an array of innovation subsets not only related to medicine but also to logistics, operations, management, as well as social and governmental elements related to humanitarian work. In this book, we take humanitarian innovation and channel its energy into a human health perspective and dissect the current state and fate of medical care and provision in conflict/crisis-afflicted areas. In looking at the delivery of health-care services in crisis situations, a proactive, yet feasible approach must be gar-nered. What we have seen is that humanitarian aid workers need not only to be properly retrofitted in responding to emergencies, but that the methods, techniques, and technologies they utilized must be passed on as well as *created* by communities within these host countries. We stress the important of the social dynamic of innova-tion—who better knows the need of conflict victims than conflict victims them-selves? While indeed knowledge transfer is critical in promoting innovation, knowledge creation is even more powerful. Crisis-afflicted communities can have opportunity to rebuild their healthcare infrastructure with the human capital and technologies gained from aid workers but also from the communities themselves. This is specifically true for the use and deployment of medical devices and tech-nologies in the field that can treat acute and chronic conditions. These devices/technologies can be created by all types of stakeholders including relief workers, conflict victims, and NGOs. Innovation observes no social boundaries. But what type of specific innovation process are we referring to?

Innovation not only refers to the products/services we create but the process itself. Of course, in resource-poor settings, the most common term used is "frugal innovations"—products and services that are aimed at contexts characterized by scarcity of capital, personnel, and infrastructure (Global Partnerships for Humanitarian Impact and Innovation 2014). This type of innovation allows humani-tarian agents to better respond to beneficiaries' needs and can drastically improve access to affordable medical technologies and increase the interventional capacities of aid and community healthcare workers in the field. Frugal innovation harnesses the distinct advantage of promoting innovation under conditions of scarcity, yet this innovation has the same interventional capacity for change as comparable to resource bountiful areas (Srinivas and Sutz 2008). Frugal innovation is executed by frugal engineering, which involves stripping out nonessential features of a device in order to lower costs while maintaining device competence and functionality (Srinivas and Sutz 2008). Although this philosophy is certainly not novel, the litera-ture exploring it remains in its early stages, and its applications to varying fields such as the one we focus on in the book, i.e., humanitarian medicine, have only begun to be explored. This type of innovation is not only derived due to limited availability of resources in order to meet the needs of low-income but also for the deployment of devices in unconventional settings such as conflict and humanitarian crises. What is unique beyond the notion of creating affordable and effective tech-nologies/devices is the thought and innovation process behind it. One does not need to have formal training in a science or engineering disciple, but rather must embrace a different type of thinking, one that focuses on feasibility, functionality, and cost in creating an effective medical device. Furthermore, in conflict and disaster settings, we cannot just simply continue to deliver goods and supplies in the same manner as the pre-disaster situation, as many times resources cannot get in through traditional logistics or supply chains; thus out-of-the-box thinking is required, and the notion of engineering unconventional technologies for an unconventional world is derived.

There are a multitude of instances where frugal innovation/engineering truly shines in relation to humanitarian innovation. For example, in post-earthquake Haiti, a pressing problem was getting medical supplies to clinics. The process of getting a shipping container full of supplies to Haiti and the materials to the right place can take anywhere from a few months to even years, and even if the materials were properly delivered, aid workers were still limited to only the materials and supplies that were brought, thus hindering the treatment capacity in the field. Such supply limitations have meant that healthcare workers in the field are often forced to resort to drastic measures such as in Haiti, where many clinics ran out of umbilical clamps, and resorted to utilizing latex gloves to tie umbilical cords. Another iconic example of the frugal innovation movement is a $25 infant warmer formally known as the "Embrace Infant Warmer," which does not require electricity (Fig. 1.5) (Banerjee 2016). This device was created to meet the needs of over 20 million low birth weight and premature babies that are born in developing countries (Banerjee 2016). It demonstrates how frugal innovation allows for more access to tools and products needed in vulnerable populations.

> *"Incubators used to cost about $30,000; they're expensive to fix and ship in, you have to use constant electric, and in a lot of environments there are rolling power outages and you can fry circuit boards."*
>
> – *Dara Dotz, Co-founder FieldReady.org*

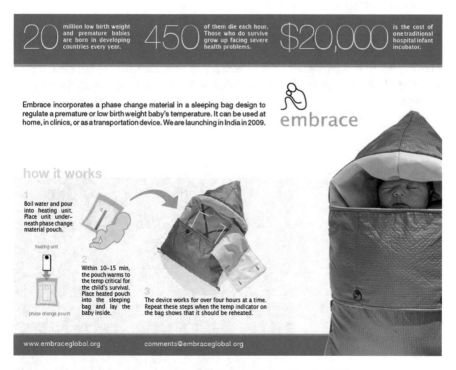

Fig. 1.5 The Embrace infant warmer device. (Embrace Infant Warmer 2016)

More than 250,000 infants in developing countries around the world have benefited from the $25 infant warmer and serve as a focal example of how frugal solutions can solve problems in simple ways (Banerjee 2016). This simple, yet powerful notion, displays the type of innovation needed in delivering medical care services in conflict, disaster, and crisis situations. But what about the role of crisis-afflicted communities in humanitarian innovation? Like we previously stated, who better knows the needs of people than the actual people themselves. This is no better exemplified than by the use of the simple sari cloth by villagers to protect themselves from cholera in countries such as Bangladesh (Huq et al. 2010). In this instance, the communities themselves worked in direct relation with researchers to ultimately identify the most frugal approach to protecting people from cholera (Fig. 1.6). Specifically, rural Bangladeshi women often pour drinks through a piece of *sari cloth* to get rid of insects and visible debris (Farmer et al. 2011). Researchers we able to determine that a sari folded into four layers can create a mesh fine enough to capture copepods, effectively removing 99% of the attached cholera bacteria in the process (Huq et al. 2010). The specific innovation process deployed is known as a "bottom-up" innovation which is an innovation process that is a subset of frugal innovation. This solution ultimately proved to be highly effective and simple. The social context of this innovation derivation is just as important as the adaptation complex. This sari cloth innovation was also adapted for deployment in humanitarian crisis scenarios such as the Haitian earthquake (Farmer et al. 2011).

We further explore the inner depths and classifications of frugal innovation processes in relation to humanitarian innovation in the next chapter, but we point out in this section that it is a vital component in transforming future interventional strategies related to healthcare access and delivery in crisis situations. But what do we mean by this? The power of frugal innovation and engineering lies in not only the novel fabrication of low-cost, value-added technologies but the *creation as well as transfer* of knowledge. This transfer and creation of knowledge is facilitated by not

Fig. 1.6 Use of sari cloth to filter cholera from water. (Huq et al. 2010)

only relief/aid workers but also community health workers that are a beacon for their communities. These individuals can facilitate the use and uptake of medical device innovations/solutions to vastly improve patient treatment and care in unconventional situations. This also includes the ability to adequately conceptualize a genuine people-centered approach to humanitarian innovation by utilizing crisis-afflicted communities as an impetus for innovation. The ability to sufficiently channel the voices of affected communities in a manner that is both desirable and possible remains one of the biggest challenges in redefining humanitarian innovation in the modern era. Although inequities in health result from the social conditions that lead to illness, the high burden of illness particularly among socially disadvantaged populations creates a focal need to make healthcare highly responsive to conflict and vulnerable population needs. This instance is precisely where community health workers come into play and can serve as an effective conduit and provider of healthcare services among these populations. What happens in this paradigm is a symbiosis whereby community health workers can adapt and react in delivering healthcare services not only in the short term but also the *long term*. By this we mean that ultimately, we want to transform and expand the knowledge of community health workers and strengthen healthcare facilities in not only buffer zones—demilitarized zones—but also regional and district-level healthcare facilities after conflict, war, or disaster crises.

Frugal engineering and innovation are vital agents in facilitating this change and enhanced treatment of conflict victims. The ability to engineer and deploy devices and equipment that are low-cost, durable, and effective can enhance the overall operating capacity of humanitarian missions and radically change the healthcare landscape in the field. From neonatal care and obstetric care to emergency trauma and surgical interventions, each of these distinct areas harbors the potential for the integration of frugal engineering initiatives, but fostering the development of feasible, high efficacy medical devices is crucial, as they must be able to sustain and operate in unconventional situations. This means that the functional design and deployment of these devices and equipment must retain its capacity to deliver it intended function yet be nimble enough to sustain repeated use in low-resource settings. Ultimately, the purpose of this is to improve human health outcomes in areas where individuals have the tendency to receive exponentially diminished care. Providing the intellectual and physical tools and resources for individuals such as community health workers as well as other healthcare workers in conflict-afflicted areas to be able to fabricate, develop, and render frugally engineered devices creates solutions for long-term integration into rebuilt healthcare systems in post-conflict settings. What this has essentially done is empowered local individuals of conflict-afflicted countries, the ability to bring back knowledge and innovation to their host countries to better develop healthcare resources.

Given that more than 60 million people are displaced from their homes by conflict and persecution in 2016, it is critical time to consider the lessons and implications of disruptive innovation (The Human in Humanitarian Innovation 2017). This involves the systematic exploitation of technologies to deliver improved or novel services in radically different ways. In the humanitarian relief paradigm, the need to redefine aid assistance and delivery in a disruptive fashion coupled with the integration of frugal innovation creates a new impetus for relief services in conflict situations. However, for humanitarian or social innovation to be functionally effective, a cross-sector and highly collaborative approach is the only option, in which the appropriate buy-in must be garnered from relevant actors to propel humanitarian and social innovation. Furthermore, the ability to deploy frugally engineered medical devices in the field can close inequalities in health among population groups, specifically the differences along multiple axes of social stratification including socioeconomic, ethnic, cultural, and gender (Orach 2009). Although the causes of inequalities in developed may be different from those in developing countries, the use of innovations and technologies that can be scaled for use to treat individuals across the spectrum means that health is restored as a fundamental tenet of the human condition. Addressing social determinants of health will yield greater and sustainable returns to existing efforts to improving global health; this includes the focal empowerment of individuals, communities, and countries (Orach 2009). Frugal engineering and innovation empowers creative solutions to problems that many humanitarian practitioners face in the field. But the true question lies in how do we actually develop and foster these initiatives to create real, feasible devices for deployment in the field? We explore the development of this thought process and its relative outputs further in our next chapter.

References

Alexander, K. A., Sanderson, C. E., Marathe, M., Lewis, B. L., Rivers, C. M., Shaman, J., Drake, J. M., et al. (2015). What factors might have led to the emergence of Ebola in West Africa? *PLoS Neglected Tropical Diseases, 9*(6), e0003652.

Annual Report: 2009. Ministry of Health Syria; (2009). Available from: http://www.gov.sy/Default.aspx?tabid=251&language=en-US. Accessed 30 Mar 2017.

Aronson, J. K. (2006). Rare diseases and orphan drugs. *British Journal of Clinical Pharmacology, 61*(3), 243–245.

Banatvala, N., & Zwi, A. B. (2000). Conflict and health: Public health and humanitarian interventions: Developing the evidence base. *BMJ: British Medical Journal, 321*(7253), 101.

Banerjee, B. (2016). Why innovate? In *Creating innovation leaders* (pp. 3–24). Cham: Springer.

Betts, A., & Bloom, L. (2014). *Humanitarian innovation: The state of the art*. New York: United Nations Office for the Coordination of Humanitarian Affairs (OCHA).

Black, R. E., Morris, S. S., & Bryce, J. (2003). Where and why are 10 million children dying every year? *The Lancet, 361*(9376), 2226–2234.

Bleck, J., & Michelitch, K. (2015). The 2012 crisis in Mali: Ongoing empirical state failure. *African Affairs, 114*(457), 598–623.

Chan, E. Y. Y. (2017). *Public health humanitarian responses to natural disasters*. Abingdon/New York: Taylor & Francis.

Chan, E. Y. Y., & Sondorp, E. (2007). Medical interventions following natural disasters: Missing out on chronic medical needs. *Asia Pacific Journal of Public Health, 19*(1_suppl), 45–51.

Chandran, A., Hyder, A. A., & Peek-Asa, C. (2010). The global burden of unintentional injuries and an agenda for progress. *Epidemiologic Reviews, 32*(1), 110–120.

Connolly, M. A., Gayer, M., Ryan, M. J., Salama, P., Spiegel, P., & Heymann, D. L. (2004). Communicable diseases in complex emergencies: Impact and challenges. *The Lancet, 364*(9449), 1974–1983.

Crisis overview 2016: Humanitarian trends and risks for 2017. 2016. *Acaps.Org*. Accessed 2 Dec 2017. https://www.acaps.org/special-report/crisis-overview-2016-humanitarian-trends-and-risks-2017

Embrace infant warmer. 2016. *T3 Middle East*. Accessed 14 Feb 2018. http://t3me.com/en/news/social-sundays-embrace-infant-warmer/

Farmer, P., Almazor, C. P., Bahnsen, E. T., Barry, D., Bazile, J., Bloom, B. R., Bose, N., et al. (2011). Meeting cholera's challenge to Haiti and the world: A joint statement on cholera prevention and care. *PLoS Neglected Tropical Diseases, 5*(5), e1145.

Fouad, F., Sparrow, A., Tarakji, A., Alameddine, M., El-Jardali, F., Coutts, A., et al. (2017). Health workers and the weaponisation of health care in Syria: A preliminary inquiry for The Lancet –American University of Beirut Commission on Syria. *Lancet*. https://doi.org/10.1016/S0140-6736(17)30741-9.

Gamble, L. (2018). Healthcare in war zones: 10 things to know. *Beckershospitalreview.Com*. Accessed 20 Feb 2018. https://www.beckershospitalreview.com/human-capital-and-risk/healthcare-in-war-zones-10-things-to-know.html

Global Partnerships for Humanitarian Impact and Innovation. 2014. International Committee of the Red Cross (ICRC) Accessed 21 Feb 2018. http://blogs.icrc.org/gphi2/wp-content/uploads/sites/96/2015/03/IMD-TEG-Global-Partnerships-Challenges-and-Opportunities-final-26-11-14.pdf

Huq, A., Yunus, M., Sohel, S. S., Bhuiya, A., Emch, M., Luby, S. P., Russek-Cohen, E., Balakrish Nair, G., Bradley Sack, R., & Colwell, R. R. (2010). Simple sari cloth filtration of water is sustainable and continues to protect villagers from cholera in Matlab, Bangladesh. *MBio, 1*(1), e00034–e00010.

Leaning, J., Spiegel, P., & Crisp, J. (2011). Public health equity in refugee situations. *Conflict and Health, 5*(1), 6.

Levy, B. S., & Sidel, V. W. (Eds.). (2008). *War and public health*. New York/Oxford: Oxford University Press.

Lewin, S., Munabi-Babigumira S., Glenton C., Daniels K., Bosch-Capblanch X., van Wyk B. E., Odgaard-Jensen J., et al. (2010). Lay health workers in primary and community health care for maternal and child health and the management of infectious diseases. *The Cochrane Library, 28*(3), 243–245.

Lipshultz, E. (2018). Securing health care in war zones. *Harvard College Global Health Review*. Accessed 16 Feb 2018. https://www.hcs.harvard.edu/hghr/online/securing-health-care-in-war-zones/

Malkki, L. H. (1996). Speechless emissaries: Refugees, humanitarianism, and dehistoricization. *Cultural anthropology, 11*(3), 377–404.

McCord, G. C., Liu, A., & Singh, P. (2013). Deployment of community health workers across rural sub-Saharan Africa: Financial considerations and operational assumptions. *Bulletin of the World Health Organization, 91*(4), 244–253b.

O'Donnell, O. (2007). Access to health care in developing countries: Breaking down demand side barriers. *Cadernos de Saúde Pública, 23*(12), 2820–2834.

Orach, C. G. (2009). *Health equity: Challenges in low income countries* (pp. S49–S51). Makerere University Medical School, Uganda: African Health Sciences.

Refugee Health. *UNHCR*. 1995. Accessed 13 Feb 2018. http://www.unhcr.org/en-us/excom/scaf/3ae68bf424/refugee-health.html

Rodrigues Santos, A. L. (2015).*Mind the gap: Designing sustainable healthcare for humanitarian aid*. Delft, Netherlands: Delft University of Technology.

Srinivas, S., & Sutz, J. (2008). Developing countries and innovation: Searching for a new analytical approach. *Technology in Society, 30*(2), 129–140.

The human in humanitarian innovation. 2017. *MISC*. Accessed 22 Feb 2018. https://miscmagazine.com/humanitarian-innovation/

Toole, M. J., & Waldman, R. J. (1993). Refugees and displaced persons: War, hunger, and public health. *JAMA, 270*(5), 600–605.

What is the Cluster Approach? | Humanitarian Response. 2017. *Humanitarian Response Info*. Accessed 7 Dec 2017. https://www.humanitarianresponse.info/en/about-clusters/what-is-the-cluster-approach

Chapter 2
Humanitarian Innovation + Medicine: Defining the Innovation Process

The term "innovation" is typically rooted in ambiguity, as it is often defined in an array of contexts, adaptations, and subject matters. Its definition ranges from the simple notion of being a new idea or method to the more robust application of enhanced solutions that meet unarticulated or existing market needs (Baregheh et al. 2009). The reality is that it is an umbrella term that describes the essence of human ingenuity and discovery, essentially an impetus for creating breakthroughs in problem-solving. This book is unique in that we garner a spectrum of distinct innovation processes and dissect them in order to divulge novel ways of improving humanitarian medicine and human health. Innovation gives way to an array of thought processes that all serve the same purpose—to enhance the way we solve problems. When it comes to the realm of human health, innovations have the capacity to enact change by quite literally having the capacity to save lives and improve quality of life. Innovation has no greater impact and application than in the scope of human health and medicine, as it has boundless potential to change the world around us. In delving deeper into the applications of innovation in human health, humanitarian medicine is perhaps one the most vital areas of application. What is unique about humanitarian aid, whether it be in the scope of disaster, relief, or general aid operations, is that it is comprehensive in relation to the human condition. These initiatives typically involve the delegation of clean water, surveillance, sanitation, emergency care, social services, preventative care, infectious disease mitigation, and clinical care. This means that the application of innovations must be targeted in scope and refined to be feasible and easily deployable in these environments. This is where bridging the gap between theory and application comes into play. While many times innovation is touted as the answer to solving humanity's most pressing problems, how exactly do we turn unconventional "out-of-the-box" ideas into feasible, conventional solutions? We explore this next in defining the four types of innovation processes that can serve as an impetus for enhancing humanitarian medicine.

K. W. Ramadurai, S. K. Bhatia, *Reimagining Innovation in Humanitarian Medicine*, SpringerBriefs in Bioengineering, https://doi.org/10.1007/978-3-030-03285-2_2

2.1   Adapting Innovation Sub-types in Humanitarian Medicine: Turning the Unconventional into Conventional

The term that is perhaps the most synonymous with innovation in low-resource settings and humanitarian crises is that of frugal innovation. Frugal innovation refers to the development of effective functional solutions to common problems with the minimal use of resources (Chavali and Ramji 2018). This type of innovation is geared toward low-resource settings when conventional solutions are too expensive, not available, nor adaptable to the particular environment or circumstances. Typically, people in these resource-constrained settings work with what they have, using affordable but effective tools, processes, and techniques to solve their problems. There are two primary forces that drive frugal innovation and contribute to the development of these tools, processes, and/or techniques. The first is from companies and/or support by organizations such as the WHO to provide accessible technologies by simplifying existing high-tech tools (Chavali and Ramji 2018). The other is from low-cost, homegrown solutions that utilize low-tech solutions to solve unmet needs (Chavali and Ramji 2018). The fact of the matter is that frugal innovation is not a novel concept by any means and is in fact a tribute to human ingenuity that has been propelled since the dawn of mankind. However, the application and refinement of frugal innovation provides many novel avenues to explore and discover. Researchers have recently broken down frugal innovation into four distinct subsets as shown in Fig. 2.1. This includes lean tools and techniques, opportunistic solutions, contextualized adaptations, and bottom-up innovations (Tran and Ravaud 2016). These sub-types are extremely vital when it comes to humanitarian medicine and enhancing the interventional capacities of relief workers/practitioners. Specifically, these innovation sub-types can serve as critical tools for enhancing human ingenuity, medical care, and relief services in the field.

Although frugal innovation is broken down into these contextual sub-types, this does not mean that these types of innovation are "siloed." In fact, these sub-types can

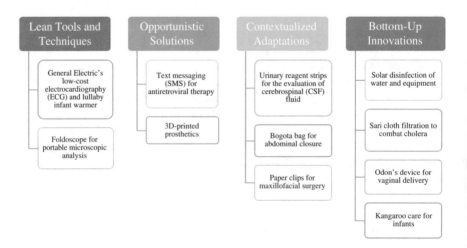

Fig. 2.1   Four sub-types of frugal innovation. (Tran and Ravaud 2016)

be utilized in conjunction with one another and even recombined to create new approaches and dynamics. Think of it as having specialized tools in a toolbelt, in which each tool serves a distinct purpose, but can be used together to fix a problem. The most important element of innovation is the recombination of previous ideas in order to create new ones. In the face of humanitarian medicine, the ability to create novel solutions that can be adapted and scaled in resource-poor and highly unconventional situations is paramount. The ability to adapt and deploy these innovation processes can serve to radically shift how we solve problems related to human health in the humanitarian field. But before we examine how to deploy these processes in redefining humanitarian medicine, we briefly define each of these sub-types.

The first process is lean tools and techniques, which refers to the simplification and adaptation of existing technologies to reduce costs and provide health innovations to the greater public (Tran and Ravaud 2016). The next type is opportunistic solution, which refers to the use of modern, widespread technologies to tackle "old problems" (Tran and Ravaud 2016). An example is the deployment of 3D printers to enhance accessibility to medical devices by allowing anyone to manufacture medical tools (Bhatia and Ramadurai 2017). The next type is contextualized adaptations, which refer to the reallocation of existing techniques, materials, and/or tools for different purposes. For example, urinary reagent strips used to evaluate cerebrospinal or synovial fluid were found to be viable diagnostic tests for meningitis in under-resourced environments at very low cost (Heckmann et al. 1996). The final type is local bottom-up innovations, which refer to original, simple ideas to obtain results not previously attainable. These innovations typically are developed in low-resource settings that challenge human ingenuity, for example, the invention of "kangaroo care" for preterm infants or solar disinfection of water to reduce diarrhea in areas where drinking water comes from waterholes not suitable for chemical treatment (Conroy et al. 1996). The next final sub-type is the bottom-up innovation, which is developed by local means and practices. For example, bicycle ambulances are a perfect alternative to car ambulances in places where cars are too costly and not adapted for the traffic density (Tran and Ravaud 2016). Although these innovation sub-types are distinct in their relative scope and application, they all follow the same fundamental innovation process schematic as depicted in Fig. 2.2. This process involves (1) specifying a problem, (2) identifying a possible solution, (3) piloting and adapting the solution, and (4) scaling the solution (Betts and Bloom 2014).

While indeed frugal innovations seek to provide solutions to common healthcare problems, they must be scientifically evaluated before widespread utilization. This is where turning the unconventional into conventional comes into play. In addition, while frugal innovations may offer effective and cheap solutions to healthcare problems in low-resource settings, there are often barriers to them being fully adopted/adapted. For example, despite the fact that flash-heating breast milk—the process of heating breast milk by using a glass jar placed in a pot of boiling water—being able to effectively reduce mother-to-child transmission of HIV infection, the process is not well implemented in African countries (Mbuya et al. 2010). This is because it requires frequent, unpractical boiling of water and because it indicates that the woman is HIV positive, exposing her to stigma in the community (Young et al. 2013). It is elements such as these that further make it imperative that these innovation sub-types be sufficiently adapted and deployed in humanitarian crisis situations.

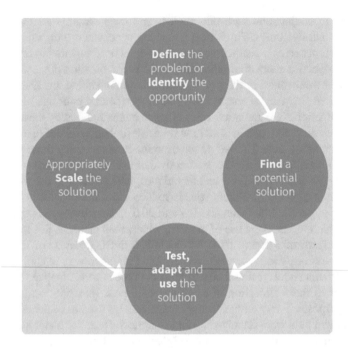

Fig. 2.2 The innovation process. (Betts and Bloom 2014)

2.2   Frugal Innovation Sub-Types in Health and Medicine

2.2.1   *Contextualized Adaptations*

Contextualized adaptations are perhaps the most easily adaptable innovation process, particularly in the humanitarian field. This simply refers to the use of existing materials, tools, or techniques and repurposing them for a novel medical use/application. This thought/innovation process is very much akin to the basis of human ingenuity and innovation. There are a plethora of devices particularly in the medical field, which can serve multiple uses and have even broader applications. One of the most prominent examples in low-resource settings is the alternative use of the Solarclave—a device consisting of a bucket, a pressure cooker, and over 100 mirrors as shown in Fig. 2.3 (Tran and Ravaud 2016). This device utilizes the sun's solar energy to heat up to temperatures of 120 °C in order to sterilize surgical equipment in the field (Tran and Ravaud 2016; Dhankher et al. 2014). Solarclave provides reliable surgical sterilization for rural clinics outside of the grid and enables healthcare workers to provide basic, life-saving services for patients. It uses locally available materials and manufacturing techniques that are already available in thousands of rural workshops across the world. Its thermodynamic efficiency allows for a small size that is easily transportable to remote clinics and is simple for one healthcare worker to set up (Dhankher et al. 2014).

Fig. 2.3 Solarclave device setup. (Chandler and MIT News Office 2018)

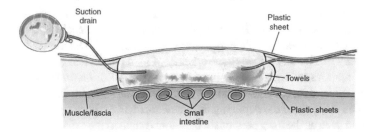

Fig. 2.4 The Bogota bag deployment schematic. (Evans and Alfred Chahine 2009)

Another example is the Bogota bag (Fig. 2.4), which is very simple, yet effective device that originated in Bogotá, Colombia. It is a sterile plastic bag used to provide temporary abdominal closure during a laparostomy. This is critical during an emergency laparotomy where the closure of the abdomen is unsafe due to the inherent risk of compartment syndrome (Ball et al. 2013). In conventional operations/cases, a 3-liter glycine bladder irrigation bag is cut open and used to cover the laparostomy (Ball et al. 2013). This allows for direct inspection of bowel in patients with ischemia, facilitates relaparotomy, improves drainage of excess fluid from the abdomen, and reduces surrounding tissue edema. It ultimately enhances fascial closure and prevents wound compromise and infection. An even wilder application of a simple plastic bag is for the prevention of acute hypothermia in preterm and low birth weight infants (Leadford et al. 2013). The effective placement of preterm/low birth weight infants inside a plastic bag at birth compared with standard thermoregulation care reduced hypothermia without resulting in hyperthermia (Leadford et al. 2013). This rather unorthodox approach defines the purest essence of contextualized adaptation in the field and represents an extremely low-cost, low-technology tool for unconventional, resource-constrained settings such as humanitarian crises.

While these applications of contextualized adaptation are remarkable, what promise does this innovation process hold for the future of innovation in humanitarian medicine? The answer to this lies in the refinement of medical interventions by humanitarian practitioners as well as forging new paths into health information and communication technologies. Many of these innovations can be classified under the umbrella term of health information technology (HIT), which involves the storage, retrieval, sharing, and use of healthcare information for communication and decision-making (Lyon et al. 2016). These technologies have become a vital element in the healthcare landscape, yet very few if any functional frameworks exist to guide their adaptation to novel applications (Lyon et al. 2016). This is where the contextualized technology adaptation process (CTAP) comes into play. The CTAP phases include (1) contextual evaluation, (2) evaluation of the unadapted technology, (3) trialing and evaluation of the adapted technology, (4) refinement and larger-scale implementation, and (5) sustainment through ongoing evaluation and system revision (Lyon et al. 2016). The CTAP process can be adapted in any environment following the schematic depicted in Fig. 2.5. The key to adapting the CTAP process in developing countries involves the contextual adaptation of one simple device—the mobile phone. These devices are highly prevalent and efficacious in developing countries and serve as a simple way to track logistical data. The use of this process can not only be vital in evaluating the efficacy of medical devices in the field but can create a mainframe where SMS mobile device applications can be utilized to enhance patient data collection and delivery of care in the field. The CTAP process also can be utilized for the deployment of new medical technologies in the humanitarian field such as the ones previously mentioned such as the Solarclave. We further dive into the development of novel technologies for humanitarian relief in Chap. 4; however, CTAP can serve as an impetus for the continued development and feasible

Fig. 2.5 The CTAP process schematic. (Lyon et al. 2016)

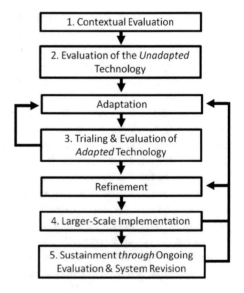

deployment of frugally engineered technologies in unconventional settings. The most important purveyors of CTAP would be the community health and aid relief workers of whom could provide instantaneous feedback with regard to how contextually adapted as well as other frugally engineered devices work in the field.

2.2.2 Bottom-Up

While indeed humanitarian response initiatives around the world vary in the scope of their response, they all have one goal—to preserve life in crisis-afflicted communities. With this common goal comes common problems, specifically that of significant resource constraints, which foster the innate adaptation complex of finding solutions in challenging environments. In creating these solutions, what is intriguing about humanitarian innovation is that these solutions are shared among a spectrum of humanitarian purveyors. NGOs, local government, and international multilateral organizations all can share in the development, deployment, and fruits of these innovation processes. It is important to note that solutions change over time and space and crisis situations and emergency response evolves over days, months, and even years (Betts et al. 2015). While we have indeed talked about the agencies behind crisis response and relief, what about the communities? How can people in these situations harbor an innovation process to preserve their health and well-being? This is where our next frugal innovation sub-type comes into play. We refer to this type of innovation that is driven by crisis-afflicted communities themselves as "bottom-up innovation" (Betts et al. 2015). Bottom-up innovation is a particularly critical yet very much overlooked component of humanitarian innovation. Typically, these types of innovations emerging from within affected communities are neglected, as individuals are perceived as vulnerable and even passive victims (Betts et al. 2015).

When it comes to the topic of humanitarianism, it often narrowly focused on improving organizational responses in humanitarian crises. While this is certainly an important facet of crisis management, the role of innovation by crisis-affected communities themselves is still a neglected topic. However, there is growing recognition of the potential for bottom-up humanitarian innovation. Organizations such as UNICEF (the United Nations Children's Fund) and UNHCR (the United Nations Refugee Agency) have worked to enhance funding to crisis-stricken communities themselves to solve pertinent problems on the ground (Betts et al. 2015). But once again, these agencies take time to fully implement structured solutions and protocols—time that is vital when it comes to the preservation of human health in crisis situations. With this in mind, there have been many cleverly developed bottom-up innovations that have served as vital agents in many humanitarian settings. For example, in Malawi and South Sudan—African countries with some of the highest maternal mortality rates in the world—communities have developed a low-tech, bottom-up innovation in the form of bicycle/motorbike ambulances (Kobusingye et al. 2005). These two-wheeled ambulances are retrofitted with a compartment for

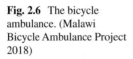

Fig. 2.6 The bicycle ambulance. (Malawi Bicycle Ambulance Project 2018)

pregnant women to sit in and be transported to community health centers/hospitals that are often out of reach to rural communities (Fig. 2.6). These ambulances can easily be deployed in crisis situations whereby expectant mothers can be easily transported to health clinics or aid stations.

So far, we have discussed frugal innovations in the form of tangible devices and technologies, but what about frugal, bottom-up innovation in the form of intellectual processes? One of the best bottom-up innovations that represent this is that of "kangaroo mother care." Rather than the creation of a device, this is a maternal/infant home care program based upon common maternal practices from local communities and villages in developing countries. The concept was formally developed in 1979 via a program launched at the Instituto Materno Infantil at San Juan de Dios Hospital in Bogota, Colombia (Ruiz-Peláez et al. 2004). Evidence has shown that kangaroo mother care (KMC) reduces mortality among babies with birth weight less than 4.4 lbs. (Vesel et al. 2015). This method is relatively simple and involves continuous skin-to-skin contact, breastfeeding support, and promotion of early hospital discharge with follow-up (Fig. 2.7) (Vesel et al. 2015). What is fascinating about this innovation is that the World Health Organization has endorsed KMC for stabilized newborns in health facilities in both high-income and low-income countries (Vesel et al. 2015; Ruiz-Peláez et al. 2004).

2.2.3 Lean Tools and Techniques

The next sub-type of frugal innovation is lean tools and techniques and is perhaps the most synonymous with the concept of frugal innovation. The reason being is that it refers to the simplification and adaptation of existing technologies to greatly reduce costs and enhance scope of application (Chavali and Ramji 2018). These consist of devices broken down to their bare bones and stripped of any accessory

Fig. 2.7 Kangaroo mother
care. (World Health
Organization: Pocket book
of hospital care for
children 2005)

*Position for Kangaroo mother care
of young infant.* Note: After wrapping
the child, cover the head with a cap to
prevent heat loss.

elements that were not vital to the device's function. This effectively results in a
device that is far more practical, cheaper, and still highly efficacious in function,
which costs as little as one-fifteenth that of their average counterparts (Chavali and
Ramji 2018). Although these technologies are low-cost versions of medical devices
used in more affluent countries, they are, in fact, their own breed. These devices are
often highly durable, portable, able to function in harsh environments, and easy to
maintain, with very cheap and accessible spare parts. In fact, lean tools and techniques that are developed for resource-poor settings are in fact sometimes so cost-
efficient that they are better than solutions used in high-income countries. For
example, engineers at Siemens developed an inexpensive CT scanner by removing
infrequently used settings and options. The new machine cut the cost of treatment
by 35% and has been readily utilized in the United States (Siemens Presents New
16-slice CT Scanner Somatom Scope 2014). In addition, two recent critical innovations were derived from this same innovation process, which includes the "Bubble
CPAP" and "Foldscope" (Figs. 2.8 and 2.9) (Chavali and Ramji 2018; Brown et al.
2013). The low-cost bubble continuous positive airway pressure (bCPAP) is a non-
invasive ventilation device for newborn babies with respiratory distress and was
developed at Rice University. The device has been deployed in district hospitals
throughout Malawi, Africa, and costs only $110 USD (Bennett et al. 2018).

Another revolutionary device created via the lean tools and techniques method is
the "Foldscope," which is an origami-based paper microscope (Fig. 2.9) (Ephraim
et al. 2015). This device is an extremely affordable microscope (less than $1 USD)
that can be utilized in identifying infectious diseases such as Malaria in rural clinics

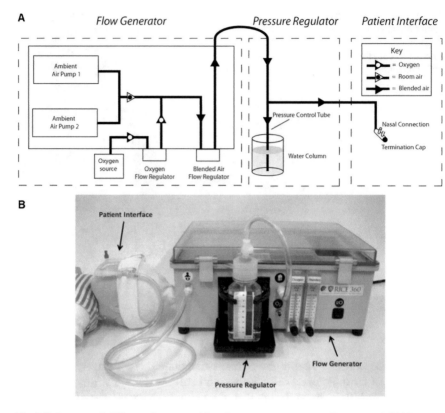

Fig. 2.8 Low-cost bubble continuous positive airway pressure system. (Brown et al. 2013)

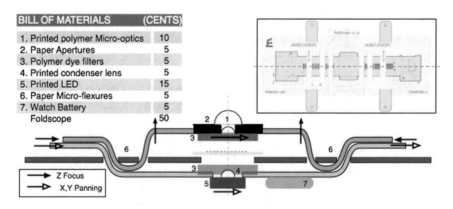

Fig. 2.9 Foldscope design schematic. (Hussey 2014)

and district-level hospitals (Ephraim et al. 2015). While indeed the device itself is impressive, what is perhaps even more exciting is the ability for lean tools and techniques to be scaled beyond simple medical devices to far greater agendas such as the entire humanitarian innovation paradigm itself. Think if we were able to take this thought process to streamline logistical support, communications, and operations in crisis situations. Once again, we find that recombination of these innovation thought processes holds the key to reimagining humanitarian medicine and innovation.

2.2.4   Opportunistic Solutions

The final sub-type is opportunistic solutions, which refer to the use of modern, cheap, and readily available technologies to tackle various problems. This is perhaps the simplest thought process, as it is simply repurposing a technology for an alternative use. This type of innovation knows no relative boundaries, as it is only limited by the innovator's capabilities and mindset. Perhaps the most vital applications of the opportunistic solution innovation process relate to that of mobile phone technologies, the Internet, and additive manufacturing or 3D printing (Tran and Ravaud 2016). These technologies can revolutionize the humanitarian innovation process and paradigm as a whole as well as radically change humanitarian medicine. Some opportunistic solutions have included the use of mobile phone SMS to improve adherence to antiretroviral therapy in Kenya. Since mobile phones are highly prevalent in developing countries around the world, Kenya pilot tested a short message service that reminded individuals to adhere to their antiretroviral treatment (ART). This ultimately improved antiretroviral treatment adherence and virologic outcomes of patients in Kenya (Mbuagbaw et al. 2013). In addition, an organization known as "GiftedMom" was developed to utilize SMS messaging to provide pregnant women and mothers enhanced access to antenatal care in Cameroon (Latchem 2018). This service has been an effective agent in reducing maternal mortality and achieving universal access to reproductive health for women in rural Cameroon (Fig. 2.10). With the rapid dissemination of information technologies in countries all around the world, the use of the Internet and SMS provides to be fruitful when it comes to humanitarian innovation and medicine. The ability to track patient's medical histories and adherence to treatment can be vital in delivering a higher standard of care.

Fig. 2.10 GiftedMom mobile application. (Gifted Mom: Cameroonian App to Promote Antenatal Care 2015)

Fig. 2.11 3D-printed prosthetics for amputees in Uganda. (Ugandan Children Get 3D Printed Prosthetics 2014)

Another opportunistic solution example is that of 3D printers. These devices hold the unique potential to effectively remodel accessibility to medical devices by allowing virtually anyone to manufacture medical tools ranging from low-cost prosthetics to spare parts of equipment (Bhatia and Ramadurai 2017). There are an array of 3D printing initiatives that have been deployed in countries throughout Africa, in which 3D printers have been utilized to print out various prosthetics ranging from arms, hands, and feet for amputees (Fig. 2.11) (Bhatia and Ramadurai 2017). These prosthetics can be fabricated on site in the field for a fraction of the cost of conventional prosthetics. Printing devices can further be retrofitted with natural, biopolymer blends that are structurally rigid and even more cost-affordable. But like any innovation process, there are really three functional parts. The first being the identification of a specific need and target market; the second is the identification of the innovation's unique attributes and selling points; and the third is achieving a cost reduction.

2.3 Disruptive Innovation: The Real Meaning

The next innovation process we define and explore in humanitarian applications is that of the "disruptive innovation." Disruptive innovation is a theory that describes technological developments which create new markets and values within an existing market, therefore displacing historic market leaders and products (Staruch et al. 2018). This innovation theory and process was first described by Harvard Business School Professor Clayton M. Christiansen in 1997 and was originally described as a business phenomenon. Since its discovery, the concept has been modelled in a plethora of novel and distinct fields including science, healthcare, economics, and engineering. In fact, the term "disruptive innovation" has become hugely popular and widely used to refer to novel innovations that shake up or "disrupt" the marketplace. While indeed the popularity of the terminology is impressive, the real meaning and application of the term have become widely misused and diluted. Disruptive innovations are in fact not breakthrough technologies that make existing products that are good better. Rather, they are innovations that make products and services more accessible and affordable to a larger population (Disruptive Innovations: Christensen Institute 2018). Furthermore, disruptive innovations must also encompass a business model that targets customers "nonconsumers," i.e., individuals who previously did not buy products or services in a given market, or "low-end consumers," i.e., the least profitable customers (Disruptive Innovations: Christensen Institute 2018). Finally, disruptive innovations must employ a network in which distributors, suppliers, and customers are each better off when the disruptive technology/innovation prospers in the future. In Fig. 2.12, we see that the different levels of innovation before an innovation can be considered destructive in nature. Oftentimes many individuals think that any new technology that is considered "innovative" or a "breakthrough" is considered to be a disruptive innovation. This is obviously false and, in reality, falls under other levels shown in Fig. 2.12—as it must embrace the qualities previously mentioned. For the sake of this book, we examine innovations employed in the humanitarian field that abide by these defining qualities.

When it comes to humanitarian medicine, disruptive innovation is very much pertinent to refugee healthcare, as traditional economic drivers are generally absent. Specifically, in order to enhance the provision of low-cost healthcare products for displaced people, understanding the sustainability complex in culmination with the economics behind disruptive innovation is vital. In the realm of humanitarian medicine, the common needs are treatment of fractures, burns, wounds, and skin diseases, as well as the monitoring for chronic diseases such as diabetes. In developing interventions for this, the deployment of treatments, devices, and data technologies in refugee camps must be durable, easy to transport, and quality-built and have a low cost. Thus, disruptive technological solutions must look beyond the delivery of

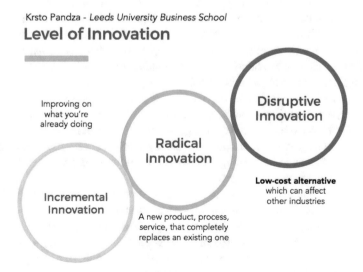

Fig. 2.12 Levels of innovation. (The Explanation of Digital Disruption in Indonesia 2017)

low-cost, high-volume products but also the introduction of new value and significant healthcare benefits. Perhaps one of the biggest areas ripe for disruption is that of medical records, documentation, and data collection in refugee encampments. While we explore the immense potential of these technologies to disrupt the humanitarian care paradigm in Chap. 4, it is important to note that these serve as true disruptive innovations. The development of cloud-based systems in conjunction with mobile phone/telecommunications networks could revolutionize refugee healthcare records. What is interesting is that most initiatives do not formally describe disruptive technology as a mechanism to improve refugee healthcare. The reason being is that there is a perpetual cycle of manufacturing unaffordable products, which do not meet the fiscal needs for low-resource settings; thus, breaking this cycle will be vital for future interventions.

Humanitarian innovation encompasses a new-market disruption, whereby we are opening a new market and focusing on users, i.e., refugees who historically lacked the money or skills to buy and use the products that are being offered, i.e., medical devices and/or technologies. When it comes to humanitarian crisis management, the reality is that the fusion of the innovation processes discussed would likely need to occur in a combinatorial manner to maximize impact and effectiveness of initiatives.

2.4   Open and Reverse Innovation + Crowdsourcing and Wikicapital: The Future of Creative Problem-Solving

When it comes to innovation, oftentimes we forget that innovation is the product of creative problem-solving. The fact is innovation is a process, not just an end product; thus, we must foster the growth and development of the innovation process in conjunction with the actual innovative product/service created. When it comes to the need of dynamic creative problem-solving tactics, there is perhaps not better area ripe for this than in global health and humanitarian relief. The innovation processes we have previously defined all serve as impetuses for creative problem-solving in the face of resource scarcity. Resource-poor setting is perhaps the most ideal catalyst for creative problem-solving and innovation—since what you got is what you got. This means that we must go beyond the conventional thought process and voyage into the unconventional. This exact paradigm is very much analogous to frugal innovation and sets the stage for two unique innovation processes we explore next—open and reverse innovation.

When frugal innovation comes to developed countries from developing countries and becomes commercially successful—such as the Siemens CT scanner—it is called reverse innovation. What is fascinating is that reverse innovation has begun to be readily adopted by many prominent firms such as GE, Siemens, and Procter & Gamble. It is no longer an innovation process that is thought to be only applicable to resource-stricken countries, but is rather considered a highly cost-effective innovation process for applications in all markets. Companies continually look to cut costs and increase profits, but in this case these cost-reduction measures not only result in enhanced fiscal outcomes for companies but also savings for consumers of their products. Reverse innovation is thought to be the aftermath consequence of frugal innovation (Immelt et al. 2009). Frugal innovation is intentionally geared toward low-income customers in developing countries, while reverse innovation turns frugal innovations into reverse innovation bringing them into the developed countries in modified form (Immelt et al. 2009). The concept of reverse innovation was first introduced as a core innovation process by Immelt et al. (2009) in Harvard Business Review. These researchers believed that reverse innovation is not optional for companies, but rather it is "as essential as oxygen" (Immelt et al. 2009). Although a reverse innovation is classified as any innovation that is adopted first in the developing world. It in fact has nothing to do with the location of the innovators, but rather the location of the consumers of the innovation (Immelt et al. 2009). Reverse innovation provides a low-price point innovation that originates from developing countries in order to satisfy consumer price points in that market, which just so happen to be appealing to consumers in developed countries (Fig. 2.13) (Govindarajan and Trimble 2012). In fact, reverse innovation holds promise in being adapted to global health systems and fostering trans-disciplinary movements. These movements seek to effectively make use of low-income country health innovations within high-income country settings.

The implications of adopting reverse innovation processes in corporate settings can indeed be a game changer for humanitarian medicine and innovation. The reason being is that international and local governmental agencies as well as NGOs can have increased access to more affordable technologies to deploy in crisis and relief

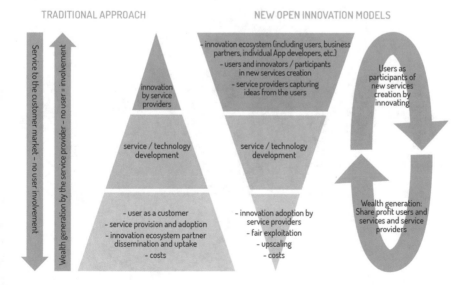

Fig. 2.13 Reverse innovation process. (Curley and Salmelin 2014)

settings. This means that humanitarian practitioners, conflicted-afflicted communities, and community health workers could indeed by more properly retrofitted with vital medical technologies at a lower barrier to entry. But once again, this sounds great in theory, but the true dissonance lies in the execution of these agendas. There are indeed many questions that must be explored with regard to innovation for frugal medical devices. These include questions such as "how many medical devices are classified as reverse and frugal innovations?" and "what can private companies and NGOs do to effectively support the development and dissemination of high-impact medical devices for humanitarian relief and developing countries in general?". These are questions that must be answered, but we explore the answers in the coming chapters.

The next type of innovation process that holds remarkable promise in humanitarian innovation and medicine is that of open innovation. This process is unique, in that it has not been readily explored in the context of low-income consumers in developing countries (Hossain 2013). It is intriguing to note that use of the open innovation process could indeed accelerate the pace of frugal innovation and reverse innovation in developing countries and in humanitarian settings. But first, what is open innovation? Open innovation is combining internal and external ideas as well as internal and external paths to market to advance the development of new technologies (Hossain 2013). Basically, it means that firms such as research institutes and businesses open up their innovation processes to be openly adapted and improved upon by external agents to propel the development of an idea or product (Fig. 2.14). Open innovation could prove to be a vital impetus for fostering humani-

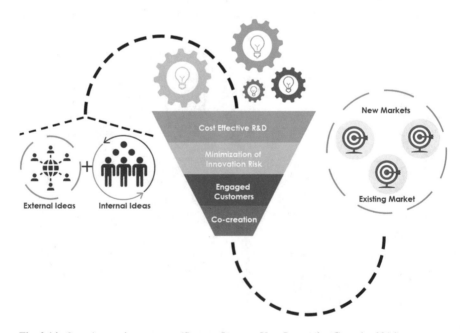

Fig. 2.14 Open innovation process. (Steps to Improve Your Innovative Capacity 2016)

tarian innovation in this day and age, as it creates a symbiosis between corporate entities and humanitarian agencies. Devices and technologies could be specifically developed for humanitarian applications and further adapted for the mainstream consumer. This effectively closes the gap between creating novel technologies for low-income populations and creating products that have effective ROIs.

What is novel in the use and deployment of open and reverse innovation is their unique propensity to stimulate bidirectional flow of knowledge and innovations between low-, middle-, and high-income countries (Syed et al. 2013). This means that no longer are developing countries or crisis-afflicted communities the recipients of one-way communication, but rather there is a proactive exchange of knowledge. This means that problems can be solved more efficiently and effectively, with the precise needs of stakeholders being met. This could not be any more critical in crisis/conflict situations, in which agencies involved must be quick to adapt and react to ever-changing problems. But this is also vital in reimagining humanitarian medicine, as practitioners can utilize these innovation processes to effectively communicate their needs in the field. This means that they can receive active engagement and feedback from sources such as Wikicapital to fabricate and develop medical devices to alleviate human suffering.

While indeed open and reverse innovation hold vast potential humanitarian innovation, where does the future novelty lie in these processes? The future of creative problem-solving and innovation lies in two unique processes: crowdsourcing and Wikicapital. Imagine having a problem and disseminating it to thousands of individuals in order to garner a plethora of unique inputs for a potential solution. This is essentially out-of-the-box thinking on another level, whereby problems can be effectively deconstructed by a massive amount of intellectual input and then an effective solution can be fabricated both quickly and efficiently. This is where Wikicapital serves as an impetus for creative problem-solving for real-world problems and is fueled by the advent of the internet and mobile phones. This open-source media platform allows for the instantaneous dissemination of knowledge on a platform that knows no physical boundaries. This effectively allows for an open forum for multifaceted user input that produces a spectrum of functional perspective that can be funneled and integrated into a single comprehensive solution. This creative problem-solving interface allows for direct communication between stakeholders on the ground and professionals both locally and internationally. Imagine if a problem in a humanitarian operation such as failed medical equipment or a logistical error, could simply be presented to an online forum of people which could instantaneously provide solutions in a live-feed fashion. This could easily be scaled for medical devices and treatment whereby devices and therapeutic inventions could be modified in real time to further suit the user/patient's needs. This means that the element of frugal innovation could be taken to another level as open-source, user feedback could further modify and enhance the device components. This would be much akin to three-dimensional printing, where 3D models can be openly adapted and improved on various platforms on the web. The ability to groupthink via Wikicapital allows for the rapid deconstruction and reconstruction of problems into effective solutions in the field. This further allows for humanitarian operators to better treat and manage conflict victims in crisis situations. We further discuss the impacts of crowdsourcing for the development of mobile and GIS tracking technologies in Chap. 4. As our world becomes increasingly connected, such interfaces as crowdsourcing and Wikicapital are paramount in not only information and knowledge transfer but the innovation process itself. Innovation is not merely a one-man job, but rather a product of collective brainpower that is constantly adapting to problems in the real world.

Wikicapital refers to the intellectual power fostered via crowdsourcing, in which crowdsourcing helps leverage the collective innovation of a large group of people to foster new ideas (Schmitz 2014). Product design has become increasingly democratized due to the advent of the Internet and social networks that more tightly connect organizations to the customers/stakeholders and markets they serve. Crowdsourcing involves the outsourcing tasks to a large group or "crowd" of people—via a platform such as the Internet (Schmitz 2014). It offers a unique way for organizations to know if products/services will succeed in the market before those products are fully deployed in the marketplace. But more importantly, it enables organizations to leverage the creativity of the masses in order to functionally improve not only their own capacity to innovate but entrepreneurs as well. Crowdsourcing gives aspiring entrepreneurs a platform to flourish and could have important implications in developing countries where many entrepreneurs have limited access to venture capital. A growing number of companies are now offering crowdsourcing platforms that enable companies to broadcast their internal problems to the world and let their audience try to solve them. For example, General Electric has partnered with the crowdsourcing platform, Quirky, to create the Inspiration Platform, a forum that invites people to innovate on technologies that GE produces (Schmitz 2014). Imagine if humanitarian operators in the field had access to a crowdsourcing service whereby they could broadcast their problems such as equipment malfunctions, medical treatment schematics, or logistical support to a dedicated community of professionals for instantaneous feedback. This could truly be a game changer for the way logistical support and problem-solution development is fostered in unconventional settings. Initiatives such as "Techfugees" have brought technologists, creatives, and entrepreneurs together to create sustainable solutions for crisis-afflicted communities and refugees around the world (6 Ways Technology Is Improving the Lives of Refugees 2016). The initiative was created by TechCrunch editor Mike Butcher that utilizes a global network of collaborators that work together via "hackathons" as well as other technological collaboration forums to find solutions to pressing problems (6 Ways Technology Is Improving the Lives of Refugees 2016). These hackathons have produced a range of solutions, such as an app for refugees to minimally share their personal details across 40 languages for nonprofits to assist in the resettlement process. Regional chapters are providing more and more opportunities around the world for people to help instigate social change for refugees—and for refugees to be part of the change.

In this chapter we explored the specific and unique capacities of various innovation processes that can be feasibly deployed in the field to actively solve problems and create feasible yet highly cost-effective solutions. In our next chapter, we explore the products of these innovation processes in relation to various medical specialties that are vital in humanitarian medicine. What we hope to create is a portfolio of current and future medical device applications that can be deployed in crisis and conflict situations to preserve human life.

References

Ball, E., Keong, N., O'Riordan, D., Shukla, C. J., Tang, T., Walsh, C., & Walsh, S. (2013). 3 emergency surgery. In *Cracking the intercollegiate general surgery FRCS viva: A revision guide* (p. 85). Boca Raton: CRC Press.

Baregheh, A., Rowley, J., & Sambrook, S. (2009). Towards a multidisciplinary definition of innovation. *Management Decision, 47*(8), 1323–1339.

Bennett, D. J., Carroll, R. W., & Kacmarek, R. M. (2018). Evaluation of a low-cost bubble CPAP system designed for resource-limited settings. *Respiratory Care, 63*(4), 395–403.

Betts, A., & Bloom, L. (2014). *Humanitarian innovation: The state of the art.* New York: United Nations Office for the Coordination of Humanitarian Affairs (OCHA).

Betts, A., Bloom, L., & Weaver, N. (2015). *Refugee innovation: Humanitarian innovation that starts with communities, Humanitarian Innovation Project.* Oxford: University of Oxford.

Bhatia, S. K., & Ramadurai, K. W. (2017). 3-Dimensional device fabrication: A bio-based materials approach. In *3D printing and bio-based materials in global health* (pp. 39–61). Cham: Springer.

Brown, J., Machen, H., Kawaza, K., Mwanza, Z., Iniguez, S., Lang, H., Gest, A., et al. (2013). A high-value, low-cost bubble continuous positive airway pressure system for low-resource settings: Technical assessment and initial case reports. *PLoS One, 8*(1), e53622.

Chandler, D. L., MIT News Office. (2018). Sterilizing with the Sun. *MIT News.* Accessed 16 June 2018. https://news.mit.edu/2013/sterilizing-with-the-sun-0226

Chavali, A. K., & Ramji, R. (Eds.). (2018). *Frugal innovation in bioengineering for the detection of infectious diseases.* Cham: Springer.

Conroy, R. M., Elmore-Meegan, M., Joyce, T., McGuigan, K. G., & Barnes, J. (1996). Solar disinfection of drinking water and diarrhoea in Maasai children: A controlled field trial. *The Lancet, 348*(9043), 1695–1697.

Curley, M., & Salmelin, B. (2014). *Open Innovation 2.0: The big picture in open innovation yearbook 2014.* Luxembourg: European Commission.

Dhankher, A., Drake, G., Haytko, J., Patel, Y., Sidoti, C., & Song, G. (2014). A solar sterilization and distillation unit for water in resource-poor settings. In *Global Humanitarian Technology Conference (GHTC), 2014 IEEE* (pp. 469–473). San Jose, CA: IEEE.

Disruptive innovations: Christensen Institute. 2018. *Christensen Institute.* Accessed 2 Sept 2018. https://www.christenseninstitute.org/disruptive-innovations/

Ephraim, R. K. D., Duah, E., Cybulski, J. S., Prakash, M., D'Ambrosio, M. V., Fletcher, D. A., Keiser, J., Andrews, J. R., & Bogoch, I. I. (2015). Diagnosis of Schistosoma haematobium infection with a mobile phone-mounted Foldscope and a reversed-lens CellScope in Ghana. *The American Journal of Tropical Medicine and Hygiene, 92*(6), 1253–1256.

Evans, S. R. T., & Alfred Chahine, A. (2009). *Surgical pitfalls: Prevention and management.* Philadelphia: Saunders/Elsevier.

Gifted Mom: Cameroonian app to promote antenatal care. 2015. *Idgconnect.com.* Accessed 12 June 2018. https://www.idgconnect.com/abstract/9810/gifted-mom-cameroonian-app-promote-antenatal-care

Govindarajan, V., & Trimble, C. (2012). *Reverse innovation: Create far from home, win everywhere.* Boston: Harvard Business Press.

Heckmann, J. G., Engelhardt, A., Druschky, A., Mück-Weymann, M., & Neundörfer, B. (1996). Urine test strips for cerebrospinal fluid diagnosis of bacterial meningitis. *Medizinische Klinik (Munich, Germany: 1983), 91*(12), 766–768.

Hossain, M. (2013). Adopting open innovation to stimulate frugal innovation and reverse innovation. *SSRN Electronic Journal, 1*, 2–6.

Hussey, M. (2014). Paper disposable microscope developed for detecting malaria. *Dezeen.* Accessed 13 June 2018. https://www.dezeen.com/2014/03/14/paper-disposable-microscope-developed-for-detecting-malaria/

Immelt, J. R., Govindarajan, V., & Trimble, C. (2009). How GE is disrupting itself. *Harvard Business Review, 87*(10), 56–65.

Kobusingye, O. C., Hyder, A. A., Bishai, D., Hicks, E. R., Mock, C., & Joshipura, M. (2005). Emergency medical systems in low-and middle-income countries: Recommendations for action. *Bulletin of the World Health Organization, 83*(8), 626–631.

Latchem, C. (2018). Health care, childcare, safe water, sanitation and hygiene. In *Open and distance non-formal education in developing countries* (pp. 121–130). Singapore: Springer.

Leadford, A. E., Warren, J. B., Manasyan, A., Chomba, E., Salas, A. A., Schelonka, R., et al. (2013). Plastic bags for prevention of hypothermia in preterm and low birth weight infants. *Pediatrics, 132*(1), e128–e134.

Lyon, A. R., Wasse, J. K., Ludwig, K., Zachry, M., Bruns, E. J., Unützer, J., & McCauley, E. (2016). The contextualized technology adaptation process (CTAP): Optimizing health information technology to improve mental health systems. *Administration and Policy in Mental Health and Mental Health Services Research, 43*(3), 394–409.

Malawi Bicycle Ambulance Project. 2018. *Bicycleambulanceproject.blogspot.com.* Accessed 12 June 2018. https://bicycleambulanceproject.blogspot.com/

Mbuagbaw, L., Van Der Kop, M. L., Lester, R. T., Thirumurthy, H., Pop-Eleches, C., Ye, C., Smieja, M., Dolovich, L., Mills, E. J., & Thabane, L. (2013). Mobile phone text messages for improving adherence to antiretroviral therapy (ART): An individual patient data meta-analysis of randomised trials. *BMJ Open, 3*(12), e003950.

Mbuya, M. N. N., Humphrey, J. H., Majo, F., Chasekwa, B., Jenkins, A., Israel-Ballard, K., Muti, M., et al. (2010). Heat treatment of expressed breast milk is a feasible option for feeding HIV-exposed, uninfected children after 6 months of age in rural Zimbabwe. *The Journal of Nutrition, 140*(8), 1481–1488.

Ruiz-Peláez, J. G., Charpak, N., & Cuervo, L. G. (2004). Kangaroo mother care, an example to follow from developing countries. *BMJ, 329*(7475), 1179–1181.

Schmitz, B. 2014. How crowdsourcing can help innovate new product design. *3Dcadworld.Com.* Accessed 19 July 2018. https://www.3dcadworld.com/crowdsourcing-can-help-innovate-new-product-design/

Siemens presents new 16-slice CT scanner somatom scope. Siemens history site – News – Werner Von Siemens is born in Lenthe. 2014. Accessed 14 June 2018. https://www.siemens.com/press/en/pressrelease/?press=/en/pressrelease/2014/healthcare/imaging-therapy-systems/him201403015.htm&content

Staruch, R., Beverly, A., Sarfo-Annin, J. K., & Rowbotham, S. (2018). Calling for the next WHO Global Health initiative: The use of disruptive innovation to meet the health care needs of displaced populations. *Journal of Global Health, 8*(1), 010303.

Steps to improve your innovative capacity. 2016. *Ideapoke.* Accessed 16 June 2018. http://blog.ideapoke.com/steps-improve-open-innovation-capacity/

Syed, S. B., Dadwal, V., & Martin, G. (2013). Reverse innovation in global health systems: Towards global innovation flow. *Global Health, 9*, 36.

The explanation of digital disruption in Indonesia. 2017. *Medium.* Accessed 27 July 2018. https://medium.com/@keilmuanmti/the-explanation-of-digital-disruption-in-indonesia-4e612a7f9be4

Tran, V.-T., & Ravaud, P. (2016). Frugal innovation in medicine for low resource settings. *BMC Medicine, 14*(1), 102.

Ugandan Children Get 3D Printed Prosthetics. 2014. *3D Printer World.* Accessed 18 June 2014. http://www.3dprinterworld.com/article/ugandan-children-get-3d-printed-prosthetics

Vesel, L., Bergh, A.-M., Kerber, K. J., Valsangkar, B., Mazia, G., Moxon, S. G., Blencowe, H., et al. (2015). Kangaroo mother care: A multi-country analysis of health system bottlenecks and potential solutions. *BMC Pregnancy and Childbirth, 15*(2), S5.

6 Ways technology is improving the lives of refugees. 2016. *Medium*. Accessed 1 Sept 2018. https://medium.com/future-crunch/6-ways-technology-is-improving-refugee-lives-5a143457f255

World Health Organization. (2005). *Pocket book of hospital care for children: Guidelines for the management of common illnesses with limited resources.* Geneva, Switzerland: WHO International.

Young, S., Leshabari, S., Arkfeld, C., Singler, J., Dantzer, E., Israel-Ballard, K., Mashio, C., Maternowska, C., & Chantry, C. (2013). Barriers and promoters of home-based pasteurization of breastmilk among HIV-infected mothers in greater Dar es Salaam, Tanzania. *Breastfeeding Medicine, 8*(3), 321–326.

Chapter 3
Frugal Medical Technologies and Adaptive Solutions: Field-Based Applications

In our previous chapter, we explore the various innovation processes that comprise frugal innovation as well as novel innovation paradigms including open and reverse innovation. Importantly, we not only define the theoretical dimensions of these innovation processes but also the functional outputs in the form of tangible technologies/devices. But while the intellectual components of these processes are critical, what does this mean for the future of humanitarian medicine and innovation? The fact of the matter is that the deployment of innovation processes in conflict and crisis situations will likely consist of an amalgam of these processes that is utilized as a catalyst for high-functioning problem-solving in the field. The reality is that crisis and conflict situations are not black and white; thus the solutions developed in the field are likely to reflect this. This is where we examine the field-based applications of these technologies and their specific capacities to preserve human life. But before we delve into these medical devices, who are these devices meant for? There are three critical stakeholders in any humanitarian healthcare operation: humanitarian practitioners (i.e., doctors, nurses, aides, relief workers), community health workers (i.e., frontline public health workers from indigenous communities), and crisis-stricken communities themselves. While the scope and capacity to utilize devices varies among these groups, nonetheless, it is vital that each one of these stakeholders be properly retrofitted with the most basic of equipment, technology, and devices. In this book we take this a step further and examine how we can not only enhance the retrofitting of humanitarian operators but also their respective problem-solving and innovation processes to create "adaptive solutions." We define these as high-utility, unconventional solutions that are derived in resource-poor settings. The reality is that while we can provide frugal devices to individuals, how do we stimulate continued innovation and the implementation of adaptive solutions on the ground? The innovation process is just as important as the device itself—a paradigm that is often overlooked.

© The Author(s), under exclusive licence to Springer Nature Switzerland AG 2019
K. W. Ramadurai, S. K. Bhatia, *Reimagining Innovation in Humanitarian Medicine*,
SpringerBriefs in Bioengineering, https://doi.org/10.1007/978-3-030-03285-2_3

3.1 Enhancing the Interventional Capacity of Community Health Workers and Crisis-Stricken Communities

The allocation of health services during humanitarian emergencies is generally non-existent or extremely frail due to the environment of violence and conflict. This weakened state is often overloaded at during times when the need for healthcare is exponentially increased. The impact of humanitarian emergencies on a population's health is severe and exacerbated by increases in food insecurity, population displacement, crowding and poor access to water and sanitation, lack of resistance to infection, the physical and psychological effects of weapons and exposure to violence, and the collapse of basic healthcare services (Van Berlaer et al. 2017). The impact of humanitarian emergencies on health workers and service provision is also extensive and includes the destruction of health facilities, infrastructure, shortages in drugs and equipment, loss of health staff, and restricted access to healthcare (Van Berlaer et al. 2017). Very often during periods of humanitarian crisis, particularly in that of resource-poor settings, we see a catastrophic failure in healthcare provisional services. So how do we overcome these deficiencies? How do we promote innovation in the face of adversity? How do we foster collective motivation and collaboration in crisis environments?

Ultimately, it comes down to a number of factors, but first we begin with community health workers (CHWs). Community health workers (CHWs) are unpaid or paid lay health workers, with a varied range of training, experience, and scope of practice (Namakula and Witter 2014). These trained individuals are often employed to mitigate against the ongoing human resource for health crises, in which 2–4 CHWs provide essential primary care at the household and community level (Gilmore et al. 2016). While the specific training and roles performed by CHWs differ across various contexts, their purpose within local healthcare systems is universal: to improve the delivery and extend the reach of primary healthcare services in a cost-effective and equitable manner (Gilmore et al. 2016). CHWs are primarily deployed in low-income and middle-income countries (LMICs), in which local governments and humanitarian organizations deploy CHW programs to increase access to care for marginalized populations (Gilmore et al. 2016). Challenges in CHW development have been well documented in countries such as Afghanistan, in which CHWs have reported difficulties with resource supply allocation, community recognition, as well as overall health systems functioning (Gilmore et al. 2016). But despite these challenges, CHWs demonstrated innovative strategies in order to functionally adapt to these challenges. CHWs are often innately tied to their communities and serve as vital representatives of the healthcare provided in the area. In order to enhance the interventional capacities of these important agents, enhanced medical device procurement must be garnered. While certainly improving the delegation of medical devices is important, the reality is that these individuals must learn how to develop these devices in resource-poor settings. This is where frugal medical devices and interventions come into play. In our next section, we explore the precise types of devices that can be developed cheaply and effectively in humanitarian crises to enhance the care delivered by CHWs in the field. The medical devices created and deployed will not only be utilized in an acute manner but also set the stage for long-term innovation processes and device development to be distributed within their local communities. The key to keeping further enhancing the treatment paradigms rendered by CHWs can also be improved by improving incentives, supervision, motivation, and continuing training (Namakula and Witter 2014). CHWs bridge the gap between humanitarian practitioners such as doctors and aid workers and crisis-stricken communities themselves.

Given how important CHWs are, how has innovation been utilized to enhance their capacities for treatment? One of the most interesting innovation initiatives has been spurred by UNICEF and is known as the "Backpack PLUS Collaboration" (UNICEF Innovation 2013). This is the product of a joint initiative between UNICEF, Save the Children, the MDG Health Alliance, and the One Million Community Health Worker Campaign and is very much the product of harboring an open innovation process paradigm. These organizations worked together to create a kit of critical tools and instruments for empowering and supporting CHWs in the field. Specifically, this toolkit was developed in order to increase their relative impact on such pressing problems such as child mortality and improve overall effectiveness and efficiency. The Backpack PLUS (BP+) contains medicines such as zinc, ORS, antibiotics, and antimalarial drugs as well as medical treatment plans and data collection devices in order to enhance the interventional and service-delivery capacities of CHWs (Figs. 3.1 and 3.2).

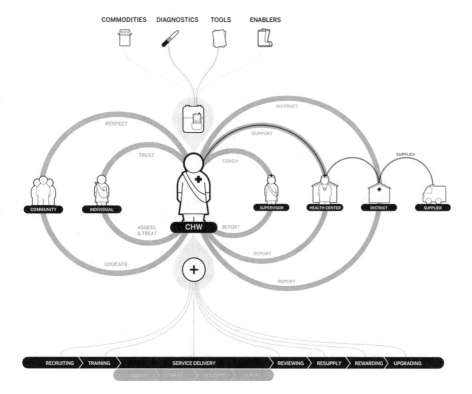

Fig. 3.1 CHW network capacity. (UNICEF Innovation 2013)

Fig. 3.2 Backpack PLUS. (UNICEF Innovation 2013)

Perhaps one of the most vital elements of CHWs is their capacity to transfer and outfit crisis-stricken communities with the tools and knowledge garnered from humanitarian practitioners. Crisis-stricken communities are oftentimes not in control of their own health but can harness simple technologies that could change their relative health outcomes. There are two critical elements to this: the ability to transfer knowledge and the ability to transfer technology. Knowledge transfer is often the most misunderstood element, as oftentimes we see a unilateral transfer of information from aid agencies to disaster-stricken communities. This means that these communities often do not have the capacity to relay information back to humanitarian operators/agencies, leaving them in the dark. This dissonance in knowledge transfer also hinders the innovation process, as many times it is the very victims of conflict that create some of the most innovative ideas to problems. The ability to dictate one's health is a fundamental human right, and the provision of frugal medical technologies can allow for the preservation of human health. The devices and applications explored in the following sections are not only reserved to enhance the interventional capacities of humanitarian practitioners/operators in the medical field for the short term but also to facilitate the long-term health outcomes and enhance the healthcare infrastructure of these communities as well.

In addition to enhancing the overall intervention treatment capacities of CHWs, what about the individuals in these crisis-afflicted communities themselves? The power of *refugee and conflict victim innovation* is perhaps one of the most overlooked elements of innovation. Oftentimes we focus on the interventional capacities of aid workers and the agencies they work for at large, but we forget that the capacity to innovate is harbored within all of us. Who better to know the needs of a community than the members of the community themselves? With regard to refugee and conflict innovation, there have been several exciting developments that seek to empower conflict victims to tackle real challenges they face. Perhaps the most promising is that of 3D printing and creating access to important supplies. One of the organizations that are at the frontier of this is that of the Jordanian organization "Refugee Open Ware" (Levin 2015). This organization was developed in order to bring important medical equipment, including prosthetic limbs, to injured survivors of the Syrian civil war (Levin 2015). The aim of the organization is to create an open-source movement for 3D printing prosthetics for refugees and disseminating innovation into the humanitarian sector. We explore the development of these innovative prosthetics in the next section, but before we delve into this, there are indeed challenges that must be observed.

Perhaps one of the biggest challenges facing refugee innovation is that of scaling. It is important to note how vast and massive many refugee and displaced-individual camps are. As depicted in Fig. 3.3, the Za'atari refugee camp located in Jordan occupies an area of more than 2 square miles and hosts more than 75,000 refugees (Jabbar and Zaza 2014). Given how large and dense these settlements are, the sufficient diffusion of innovations to meet the needs of their targeted stakeholders is vital. This is where the power of crowdsourcing and open innovation can come into play. Refugee camps often harbor an immense array of intellectual talent and human capital. Harnessing the power of engineers, academics, and other displaced

Fig. 3.3 Aerial view of the Za'atari Refugee Camp. (Behind the Fences of Jordan's Za'atari Refugee Camp 2016)

members of society is indeed vital to not only enhancing interventions but ensuring the sustainability of these interventions to help people in need.

When it comes to deploying the innovation processes we have explored in humanitarian medicine, there are an array of various medical disciplines that have and can utilize it. In the coming sections, we explore how basic innovation processes were utilized to harbor powerful, cost-efficient, yet highly efficacious medical innovations that have the distinct potential of enhancing humanitarian medicine.

3.2 Scaling Adaptive Solutions in the Humanitarian Field

The initial impetus for the development of low-cost, quality innovations in developing countries is the ability to satisfy acute need in the local setting and not the potential for export back to rich countries—and thus the deployment of the reverse innovation paradigm. But yet the need for low-cost, high-quality innovations for medicine, in essence, truly knows no borders. When it comes to general surgery and surgical care, surgical tools and instruments are perhaps one of the biggest areas that have benefited from the collective creative intelligence of physicians practicing in resource-poor settings. There are an array of surgical innovations—that are now deployed as standard therapies worldwide—whose origins are from developing countries. But what is the capacity for these technologies/innovations to be deployed in surgical interventions in the humanitarian field? This, of course, we explore in this section and further delve into the applications and derivations of humanitarian field innovations ranging from 3D printing prosthetics for refugees to the use of microfluidic paper-based analytical devices for the quick and efficient diagnosis of chronic and pathogen-derived illness. The key to the practical implementation of these innovations is not only the ease of their relative innovation processes but also their practicality and low-cost nature that make them easy to utilize in resource-poor settings. This not only enhances the interventional capacity of the humanitarian practitioner but also that of displaced/conflict victims and refugees of whom can utilize these devices. The further development of these devices via "refugee innovation" processes could indeed serve as a focal paradigm shift in the way medical care is deployed in relief settings.

3.2.1 Surgical Care and Prosthetics

When it comes to refugee and conflict victim displacement in humanitarian crises, oftentimes these individuals are resettled in remote areas. Many refugee camps in countries such as Uganda, Bangladesh, Syria, Jordan, and Kenya are often far from urban areas and settlements. Given the remoteness of these areas, technology and innovation can indeed be vital in addressing some of the enormous issues facing the ever-growing refugee populations. But deploying technologies in these remote areas requires highly adaptable solutions that can function in often austere environments. What is fascinating is that these resource-poor settings often spur the most interesting innovative processes related to medicine—particularly orthopedics and prosthetics. We examine not only the medical devices/applications themselves but also the innovation processes behind them. In particular we examine the intersection of frugal, open, and crowdsourcing innovation in developing 3D-printed prosthetics as well as the process of reverse innovation in the development of the Arbutus Drill Cover System.

Innovation is no stranger to the creation of prosthetics, as the creation of the highly effective and cost-efficient Jaipur leg in India is perhaps one of the most cited examples of frugal innovation today. But what about the future? The future of prosthetics lies in the provision of custom-made prosthetics via 3D printing to people living in refugee camps. But the innovation is not simply harbored via this creative notion of fabricated custom, low-cost prosthetics, but the cultivation of an innovation ecosystem to accompany it. This is perhaps no better emulated than by the organization known as Refugee Open Ware (ROW). ROW works to create and develop 3D-printed prosthetic limbs for Syrians and refugees that could not afford to receive conventional prosthetics. Specifically, they are working on deploying the "E-Nable hand prostheses" (Fig. 3.4) to refugee camps in Jordan, where thousands of Syrians escaped from the ongoing civil war in one of the most dramatic humanitarian disasters in history (Sher 2015). Given that a high number of injuries caused by the conflict require amputation, ROW is looking to help those affected by producing 3D printed prosthetics faster and cheaper than via conventional methods. Estimates of the number of amputations caused by the Syrian conflict are as high as 200,000 individuals, and in countries such as Jordan, there is a dire need for functional, low-cost, and rapidly produced prosthetics (Sher 2015). This is even more critical for pediatric applications as there is often a dire need for frequent prosthetic replacements for the many child amputees as they grow. ROW has thus far created 3D-printed prosthetic hands for a Yemeni child and a Jordanian boy and have further

Fig. 3.4 The E-Nable hand prosthetic. (Sher 2015)

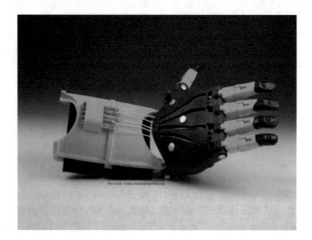

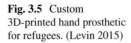

Fig. 3.5 Custom
3D-printed hand prosthetic
for refugees. (Levin 2015)

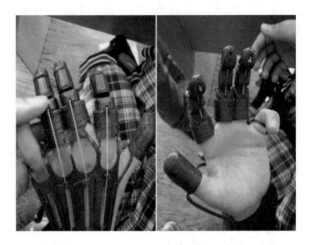

developed a finger replacement model for a 13-year-old Syrian refugee who lost several fingers (Figs. 3.4 and 3.5).

What is perhaps the fascinating element in regard to ROW's operation is not only the frugal innovation it proposes in the low-cost prosthetic, but the frugal and open ecosystem they wish to establish within the refugee camps themselves as stated by ROW founder, Dave Levin:

> We aim to build a digital fabrication lab (fablab) in Za'atari Camp, on the Jordanian border with Syria. Za'atari hosts 85,000 Syrian refugees who have built a DIY informal settlement with little more than basic tools...this lab will offer educational programs, vocational training, business development and psychological treatment through interactive art. We expect the lab to grow organically into an Open Innovation Center: a place to crowdsource, co-create and test solutions to moonshot humanitarian innovation challenges.

This approach to custom prosthetics is rooted in the tenets of frugal (specifically the opportunistic innovation process) and open innovation, whereby an opportunity is identified and the associated barriers to entry for a custom prosthetic are overcome by lowering the per-unit cost while delivering the same value, with open and direct user feedback. Furthermore, ROW is in the process of developing a human-centered design approach to create a new open-source 3D-printed prosthetic that will be culturally appropriate in the Middle East. What we see here is that the innovations are not only developed for humanitarian applications, but further utilized in more conventional applications (Fig. 3.6).

While indeed there is much novelty in this approach, perhaps the most exciting element of this venture is the idea of training those affected by a humanitarian crisis to use certain technologies to create tangible solutions they require. An individual who has grown up in a conflict region could be equipped with the skills and knowledge that can help rebuild their community and create new industries and jobs, for a whole generation. This knowledge transfer and the use of technology and human capital to foster it is what is vital in creating sustainable innovation. The key to any innovation is the human element, and the facilitation of education and skill development for refugees and conflict victims serves to not only serve them in the short

Fig. 3.6 Digital rendering of custom hand prosthetic. (Levin 2015)

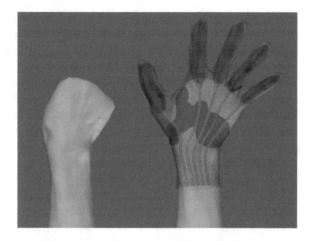

term, but the long term. Given the immense benefits of this, a network of humanitarian actors has set up a base in Jordan to train refugees in what is known as "open source tech" (Levin 2015). The goal of this initiative is for the refugees themselves to harbor the intellectual capital and resources to utilize these skills to address issues rising from conflict areas such as Syria. The key is the open-source nature of the technology associated with 3D printing including mobile computer-aided designs (CAD) which are crowdsourced blueprints for an array of products including medical devices and tools (Bhatia and Ramadurai 2017). The simultaneous deployment of human capital in conjunction with open-source technology such as 3D printing creates a paradigm shift whereby no longer are humanitarian operators conventional agents such as aid workers, but, rather, the conflict victims and refugees themselves. This creates an instantaneous feedback loop in which designs and applications of prosthetics can be modified based on infield feedback but also input from online communities that can further adapt and configure prosthetic designs. This ultimately creates a lower barrier to entry to the deployment of prosthetics to conflict victims in the field. The use of open and frugal innovation processes can lower the price and functionality of the devices for practical deployment.

The intersection of open and frugal innovations in humanitarian medicine is profound, but what about reverse innovation? This fascinating innovation concept, while in its infancy, is actually a highly practical process that seeks to radically change the fiscal barriers to entry not only in developing countries but also developed countries as well. In the field of orthopedics, there are quite a few examples of devices and applications that were fostered in developing countries and can be vital not only for humanitarian medicine, but medicine in general. One of the cleverest cost-saving innovations has been utilized in Malawi and Uganda to reduce the medical device equipment costs associated with orthopedic surgery. It is certainly unconventional, but if it reduces costs while maintaining practical functionality, it is an innovation worth deploying! This innovation is the application of a sterile bag to house a standard power hardware store drill, more formally known as the "Arbutus Drill Cover System" (Darzi 2017). This simple innovation costs less than $50 and can turn any standard drill that costs less than $100 into a functional surgical drill. This compares to high-cost surgical drills than can cost more than $15,000, creating a significant cost barrier for humanitarian applications (Darzi 2017). The drill is constructed with heat-resistant materials, allowing it to be sterilized, in which only the wrapping needs sterilizing—thus saving money on the expensive heat-resistant materials needed for conventional surgical drill fabrication. As shown in Fig. 3.7, the drill cover is a waterproof, autoclavable surgical bag that connects to the drill's mechanics via a sealed bearing mechanism. Furthermore, this drill was found to be just as safe and effective as expensive versions.

Surgical drills cost tens of thousands of dollars, yet from a mechanical standpoint, they are no different from ordinary household drills. But why the massive price disparity? This is because a surgical drill must be sterilized after each use, in which the drill must be capable of withstanding 30 min submerged in an autoclave—a pressurized steam sterilizer heated to 250° Fahrenheit (Darzi 2017). The drill cover seals a regular household drill so that only the drill bit and attachment need to be sterilized with the removable cover after each use. But where does reverse innovation come into play? Given the significant cost savings of this frugal innovation, the opportunity to scale it from Malawi and Uganda to more established healthcare systems such as England's National Health Service (NHS) was intriguing. Given that the average surgical drill is in operation for approximately 5–7 years, the turnover cost of replacing all those needed in the NHS is estimated at more than $160 million (Darzi 2017). With the frugal drill cover innovation, the cost of replacing these drills would only amount to a few hundreds of thousands of dollars—a massive cost savings for the system. Given this, the adapted drill has been successfully used to treat 30,000 patients so far in 50 hospitals across more than 10 different countries.

What we see with the example of the drill cover system is not only the frugal and reverse innovation processes in play, but the important theme of creating bilateral knowledge transfer. By this we mean that no longer are innovations derived in

Fig. 3.7 Low-cost sterile orthopedic drill cover. (Drilling to the Problem: Low-Cost Sterile Drill Covers for Surgery 2014)

developed countries for applications for developing countries, but vice versa. Another fascinating medical device innovation that is similar in scope and application is the use of mosquito net mesh for inguinal hernia repair: at a cost of more than $125 per patient, commercially produced mesh is generally deemed as unaffordable for the majority of hernia patients living in LMICs as well as refugee and internal displacement camps (Cotton et al. 2014). Faced with this cost barrier coupled with the high demand for hernia repair, a contextualized adaptation innovation process was deployed to create a novel solution, i.e., mosquito net mesh for hernia repair. This involved repurposing a device—in this case, a mosquito net—for a novel application, i.e., hernia repair (Fig. 3.8). The cost of the mosquito net mesh is only $1, yielding a savings of more than 99% when compared to conventional mesh (Löfgren et al. 2016). Furthermore, after the use of the contextualized adaptation process, reverse innovation was applied to scale this medical device application from developing to developed countries. The use of mosquito net for hernia repair traces its roots to South Africa, where surgeons utilized cheap, readily available sterilized mosquito net mesh to repair hernias instead of expensive commercial mesh (Cotton et al. 2014). This frugal version and application is just as effective and safe as commercial mesh, but the cost difference is more than $120 per piece of hernia mesh (Löfgren et al. 2016).

With regard to the reverse innovation processes behind this innovation, hernia repair is one of the most common surgical procedures in the United States and the United Kingdom. The procedure generally involves the use of a piece of surgical mesh stitched into the abdominal wall to strengthen it, with hundreds of thousands of these operations being performed annually. Given the per-unit cost of the commercial mesh, the use of sterilized mosquito mesh could indeed be scaled to level whereby healthcare systems could be saving tens of millions of dollars related to the operational cost of just this one procedure. With regard to the implications of innovations such as this as well as other contextualized adaptations in humanitarian medicine, there are truly no limits to scope, application, and importance. When it comes to refugee and internal displacement camps, utilizing basic resources such as mosquito nets—which are low cost and readily available—is absolutely vital for

Fig. 3.8 Mosquito net for hernia repair. (Operation Hernia 2018)

breaking down the barriers to care in these resource-stricken settings. This can allow for the adequate delegation of vital surgical procedures in humanitarian settings, thus enhancing the care and surgical service provision paradigm for both patients and physicians in the field. The ability to develop contextualized adaptations in the surgical field is vital to not only tearing down the barriers for accessible care for conflict victims, but further enhancing the delegation of surgical care and delivery on a global scale. We further see this trend in our next segment that explores the development of low-cost medical interventions for humanitarian applications related to neonatal and maternal conditions.

3.2.2 Maternal Conditions

In the next segment, we examine the applications of medical device innovations and humanitarian medicine in relation to an essential medical area—neonatal and maternal conditions. One of the most common areas for medical treatment related to refugee and conflict victim displacement is that of neonatal and maternal conditions.

Worldwide, more than 13 million births each year face serious complications, in which more than 800 women die each day from preventable causes related to pregnancy and childbirth (Schvartzman et al. 2018). In crisis and conflict settings, the susceptibility to fatal maternal and neonatal conditions exponentially increases due to a lack of access to basic medical care and services. The importance of adequate treatment of patients with the conditions we discuss in this segment is paramount in not only promoting mother-baby health, but women's health equity in general. Crisis and conflict situations place immense stress and burden on expectant mothers, and the inability to have access to sufficient maternal care could have dire consequences. Maternal health conditions including postpartum hemorrhage and preeclampsia can lead to severe pregnancy complications resulting in perinatal morbidity and mortality. The ability for practitioners to deliver care to circumvent these conditions is vital, but in order to do so, low-cost, highly efficacious medical innovations must be developed.

The first innovation we explore is that of the "Odón device." This device is a low-cost technological innovation that facilitates operative vaginal delivery and is designed to minimize trauma to both the mother and baby (Schvartzman et al. 2018). These features combined make it a potentially revolutionary development in obstetrics, particularly for improving intrapartum care and reducing maternal and perinatal morbidity and mortality in humanitarian settings. The low-cost device consists of a plastic sleeve that is inflated around the baby's head and is used to gently pull and ease the head of the infant through the birth canal as shown in Fig. 3.9. (Schvartzman et al. 2018; McNeil 2017). The use of forceps and other mechanical devices in the extraction of a baby in a difficult delivery can cause internal bleeding in the mother or may result in injuries to the baby's head, neck, or spine (Schvartzman et al. 2018; McNeil 2017). The Odón device has the potential to allow for vaginal delivery in complicated pregnancies where cesarean section, the use of

1

The inserter is applied on the head of the baby. A soft plastic bell assures perfect adaptation to the fetal head and prevents damage.

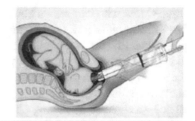

2

The inserter progressively positions the Odón device around the head of the baby. Positioning occurs as the inserter gently produces the sliding of the two surfaces of the folded sleeve along the birth canal and around the baby's head.

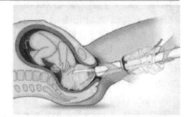

3

When the Odón device is properly positioned, a marker on the insertion handle become clearly visible in the reading window. A minimal and self-limited amount of air is pumped into an air chamber in the inner surface.

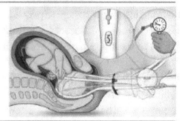

4

This produces a secure grasp around the head of the baby that fixes the inner surface and allows for traction. The inserter is removed.

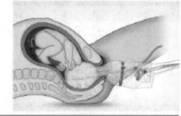

5

The head is delivered taking advantages of the sliding effect of the two surfaces of the folded sleeve. Lubrication of the surfaces further facilitates the extraction process. If needed, traction can be applied up to 19 kg (which is equivalent to the force applied with the metal vacuum extractor).

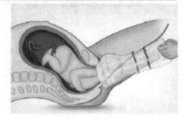

Fig. 3.9 Use of the Odón device for vaginal delivery. (Schvartzman et al. 2018)

forceps, or the use of a ventouse vacuum would be utilized. By reducing contact between the baby's skull and the birth canal, the risk of infection is also reduced.

What is fascinating about the Odón device—and the reason we explore it in this section—is the innovation process deployed behind it. The device was not developed by a physician or engineer, but rather a car mechanic from Argentina named

Jorge Odón, who had seen a YouTube video showing how to retrieve a loose cork from inside an empty wine bottle (McNeil 2017). This was done by inserting a plastic bag into the bottle, inflating the bag once the cork was inside, and then pulling out the inflated bag together with the cork (McNeil 2017). Odón conceived of the use of this same technique that evening in bed and spoke with a local obstetrician who believed the idea held merit for further development. The prototype of the device was created by sewing a sleeve onto a cloth bag and was tested using a doll inserted into a glass jar to simulate the use of the device in the delivery process. This is an excellent example of both a lean technique and opportunistic solution innovation process being deployed that ultimately took to fruition as a reverse innovation as well. The device's $50 cost will likely be further reduced for use in LMICs as organizations including the World Health Organization have fully endorsed the product for use in the field. The device is a fantastic example of how innovation can be fostered by anyone regardless of specialized expertise and background.

The next device we explore is a low-cost uterine balloon tamponade to control postpartum hemorrhage. Specifically, we explore an evidence-based package called Every Second Matters for Mothers and Babies-Uterine Balloon Tamponade (ESM-UBT) to treat postpartum hemorrhage (PPH) as shown in Fig. 3.10. The ESM-UBT is a simple innovation that consists of a condom fastened to a French Foley catheter by string (Makin et al. 2018). A condom is utilized as it provides a low-pressure system which can accommodate a high volume and conforms to the space it is inflated within (Makin et al. 2018). PPH accounts for approximately 127,000 deaths worldwide annually, 99% of which occur in low-resource settings (Makin et al. 2018). PPH is utilized when first-line treatments for postpartum hemorrhage fail to sufficiently control bleeding. In order to stop the bleeding, one potential option is to

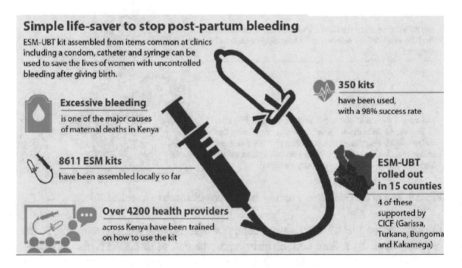

Simple life-saver to stop post-partum bleeding

ESM-UBT kit assembled from items common at clinics including a condom, catheter and syringe can be used to save the lives of women with uncontrolled bleeding after giving birth.

Excessive bleeding is one of the major causes of maternal deaths in Kenya

350 kits have been used, with a 98% success rate

8611 ESM kits have been assembled locally so far

Over 4200 health providers across Kenya have been trained on how to use the kit

ESM-UBT rolled out in 15 counties

4 of these supported by CICF (Garissa, Turkana, Bungoma and Kakamega)

Fig. 3.10 ESM-UBT Kit. (Scaling Up Life Saving Innovations for Mothers and Newborns 2018)

insert a balloon tamponade in the uterus. While indeed uterine balloon tamponades have been utilized in developed countries for many years, the associated costs and lack of access to equipment and training have limited their use in low-resource and humanitarian settings (Makin et al. 2018). The ESM-UBT consists of a condom that is tied to a Foley catheter and then inflated with clean water through a syringe and one-way valve. Using readily available materials, it is especially suitable for use in remote low-resource settings and offers the same efficacy of traditional commercial UBTs. The uterine balloon tamponade can be made available to all women who give birth in a variety of settings and costs only a few dollars per kit. The ESM-UBT is indeed a vital medical device in the humanitarian practitioner's toolkit and further serves as a gender-equity promoting agent, via its capacity to preserve women's health and fully prevent preventable death from childbirth.

The final innovation we explore in the treatment of neonatal and maternal conditions in the humanitarian field is that of the Congo Red Dot (CRD) test to diagnose preeclampsia in unconventional, resource-poor settings. This diagnostic device requires minimal equipment, is extremely cost-effective, and can be deployed by almost anyone. This innovation is not only novel in the scope of its application, but the disruptive innovation paradigm it represents related to mobile health—otherwise known as mHealth (Jonas et al. 2015). In recent years, the concept and applications of mHealth have dramatically expanded in scope and scale beyond simply electronic health (eHealth). Specifically, it now refers to the "use of mobile computing and communication technologies in health care and public health" (Jonas et al. 2015). What is fascinating about the advent of the smartphone is its relatively untapped potential. Specifically, these devices are often equipped or retrofitted with powerful embedded sensors, processors, and applications that have largely been underutilized. Given this untapped potential, developing smartphone-based diagnostics represents the new front of disruptive innovation in mHealth. The disruption, however, is not emulated in the simple duplication of existing tests but the creation of new tests that sufficiently utilize the defining molecular characteristics of a disease pathology to lead to its acute detection and diagnosis. In this case we explore the deployment of the CRD test as an mHealth application to detect and diagnose preeclampsia in female refugees and internally displaced conflict victims.

Preeclampsia affects 1 out of every 20 women and is a completely treatable, yet highly prevalent pregnancy-related disease that is responsible for maternal and fetal morbidity and mortality around the world. In settings with limited medical resources, this condition is responsible for the preventable deaths of thousands of women each year due to the lack of sufficient screening in healthcare facilities. Preeclampsia is typically diagnosed based on the symptoms of hypertension and proteinuria occurring in pregnancy after 20 weeks of gestation (Jonas et al. 2015). Given the preventable nature of morbidity with the detection of preeclampsia, there is indeed a clear need for a new diagnostic testing paradigm specifically developed for humanitarian settings that is easy to use. The Congo Red Dot (CRD) test is a molecular diagnostic test for preeclampsia—developed by researchers at the Ohio State University and

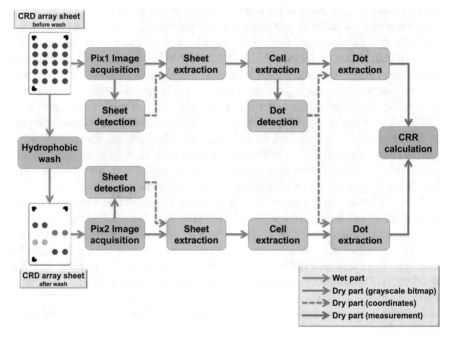

Fig. 3.11 CRD test and processing schematic (Jonas et al. 2015)

Nationwide Children's Hospital—that works via the bonding of amyloidophilic dye Congo red to misfolded proteins in the urine and is more than 85% effective at diagnosing the condition (Fig. 3.11) (Jonas et al. 2015). These misfolded proteins represent a molecular feature that is directly proportional with disease severity. This test can be coupled with smartphone-based imaging in order to significantly shorten the processing and diagnostic timeframe. The interventional capacity for device utilization is also a key feature, as it enables minimally trained personnel to effectively diagnose preeclampsia in the field.

According to the United Nations Children's Fund (UNICEF), approximately 80% of women in developing countries receive antenatal care (ANC) from a skilled health provider at least once during the course of their pregnancy (Jonas et al. 2015). Given the extremely low frequency of ANC visits, this makes the detection of preeclampsia extremely difficult, as the failure to measure blood pressure and proteinuria at each ANC visit is indeed a missed opportunity to diagnose preeclampsia. However, when it comes to humanitarian, crisis, and conflict situations, ANC is essentially eliminated, increasing the likelihood of undiagnosed preeclampsia exponentially. Due to its simplicity and the low cost of required materials, the CRD test has the potential to fill this gap for diagnosing preeclampsia in humanitarian settings and developing countries in general.

3.2.3 Infectious Diseases

When it comes to monitoring human health in humanitarian crises, there is perhaps no greater threat than that of infectious diseases. Infectious diseases remain a significant contributor to the global burden of disease in LMICs. This includes an array of diseases ranging from tuberculosis and malaria to diarrheal diseases and lower respiratory infections. These conditions kill more than 11 million people in LMICs each year and are, in many times, entirely preventable (Black et al. 2016). In looking closer, the burden of disease often falls disproportionately on vulnerable parts of the population. Specifically, more than 95% of deaths from respiratory infections and 98% of deaths from diarrheal diseases occur in LMICs and plague-specific demographics such as the elderly and children under the age of 5 (Black et al. 2016). Furthermore, war, conflict, and natural disasters open the door for disease epidemics and contribute to significant loss of life. The outbreaks of cholera during the Haitian earthquake and the recent diphtheria crisis in the Kutupalong refugee camp that is home to thousands of displaced Rohingya refugees in Bangladesh are prime examples of infectious disease that knows no bounds. Infectious diseases often plague conflict and internally displaced victims as they do not have sufficient access to infrastructure such as internal plumbing, clean water, hygiene stations, and medical clinics. Thus, the efficient diagnosis and treatment of diseases such as cholera, dysentery, diphtheria, typhoid, pneumonia, tuberculosis, and other diseases is vital in preventing death and suffering (Black et al. 2016). Innovations in diagnostic technologies are vital in the diagnosis, monitoring, and expanded coverage of lifesaving treatments. This is particularly important in pediatric applications, as children under the age of 5 are extremely vulnerable and susceptible to an array of infectious diseases leading to death. The best treatment of infectious disease is prevention—if we are able to prevent the onset of illness, this radically simplifies the amount of resources and capital deployed in the accompanying treatment and care paradigm. But once again, how does the innovation process play a role in this, and what are some examples of highly efficacious and simplistic disease treatment/prevention innovations?

The first innovation we explore is that of chlorhexidine—a perfect example of a contextualized adaptation innovation. Chlorhexidine, a low-cost antiseptic, has recently been discovered as a highly effective, easy-to-use application in umbilical cord care to prevent life-threatening neonatal infections in newborns born in unconventional, unsanitary conditions. On a global level, neonatal infections account for over 1 million newborn deaths annually, many of which could be prevented with simple interventions such as chlorhexidine (Gathwala et al. 2013). The reason why umbilical cord treatment is so vital is that after it is cut, it is prone to bacterial infections that can travel into the bloodstream and cause acute newborn sepsis and death (Gathwala et al. 2013). Chlorhexidine digluconate comprises of various forms and has been used for nearly 50 years with applications across a broad range of veterinary, dental, and medical indications (Gathwala et al. 2013). With regard to umbilical cord treatment, chlorhexidine costs less than $1 per application and is a clinically proven intervention that has been reformulated as 7.1% chlorhexidine digluconate

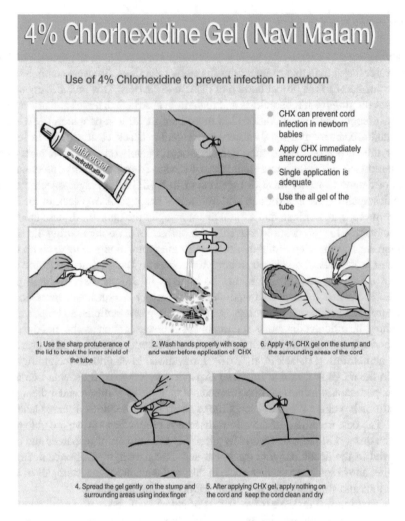

Fig. 3.12 Chlorohexidine application schematic. (Demand Generation I-Kit for Underutilized, Life Saving Commodities 2018)

solely for umbilical cord use (Gathwala et al. 2013). Chlorhexidine has no known toxicity risks and can rapidly reduce newborn deaths. Furthermore, it has a long shelf life and can be utilized with minimal training and no equipment. Thus, it is extremely efficacious for the use not only in conventional environments such as hospitals and healthcare centers but also in the field in humanitarian settings (Figs. 3.12 and 3.13).

Contextualized adaptations and innovations such as chlorohexidine show that the innovation process does not have to reinvent the wheel, but rather utilize it for novel purposes. In further developing this analogy when it comes to innovation in

Fig. 3.13 Chlorhexidine application to a newborn in India. (Segre 2013)

humanitarian medicine, we can further examine how we can create different types of "wheels" that further propel innovation paradigms. The next innovation we explore is that of microfluidic paper-based analytical devices otherwise known as µPADs (Lam et al. 2017; Martinez et al. 2009). These diagnostic devices are a new class of point-of-care diagnostic devices that are inexpensive, easy to use, and designed specifically for use in developing countries and unconventional, resource-poor settings and applications (Martinez et al. 2009). According to the World Health Organization, diagnostic devices for developing countries should be "ASSURED" (Sher et al. 2017). This acronym stands for affordable, sensitive, specific, user-friendly, rapid and robust, equipment free, and deliverable to end users (Sher et al. 2017). This simple acronym is a powerful set of guidelines that encompasses the true ability for devices to properly work in various environments. ASSURED devices should be the benchmark for innovation in humanitarian medicine and for devices utilized in any environment regardless of economic classification. µPADs are a novel, innovative platform designed for ASSURED diagnostic assays, in which they combine the capabilities of conventional microfluidic devices with the simplicity of diagnostic strip tests (Sher et al. 2017; Martinez et al. 2009). What is fascinating is that this represents a combinatorial innovation—involving the fusion of previous device components into a novel, comprehensive innovation. For µPADs this is represented in the mix of the simplicity of a diagnostic test strip with that of the functional capacity of a microfluidic device. The resulting innovation is that µPADs hold promise in providing rapid, inexpensive bioanalyses that require minimal volumes of bodily fluid such as blood or saliva. Furthermore, these devices can operate with limited equipment or power since fluid movement in µPADs is controlled via capillary action (Lam et al. 2017; Martinez et al. 2009).

µPADs represent an inexpensive point-of-care (POC) device that can provide rapid and accurate results in the field (Syedmoradi and Gomez 2017). A µPAD consists of paper, which is hydrophilic in nature and allows hydrophobic demarcations to be made with various polymers. Paper is one of the promising materials for making bioanalytical devices such as µPADs since it is inexpensive, widely available, and hydrophilic, which allows solutions to flow through it via capillary action. The innovation in these diagnostic devices is not only present in their efficacy and efficiency but also their adaptability complex. These devices can be utilized in conjunction with camera-enabled phones in order to collect data and images that can be transmitted to centralized laboratories in order to garner results in real time (Fig. 3.14) (Syedmoradi and Gomez 2017).

We can see that the efficacy of this device is promising, but what is the interventional capacity for this device in humanitarian medicine? µPAD biosensing platforms have been developed to detect various infectious diseases including human immunodeficiency virus, *E. coli*, tuberculosis, pneumonia, and *Staphylococcus aureus* (Lam et al. 2017; Martinez et al. 2009). These devices often utilize a modified gold nanoparticle solution—with specific pathogen or biomarker recognition elements—that is transferred to cellulose paper. When placed on the cellulose paper, the bacterial samples induce nanoparticle aggregates that can be detected. The

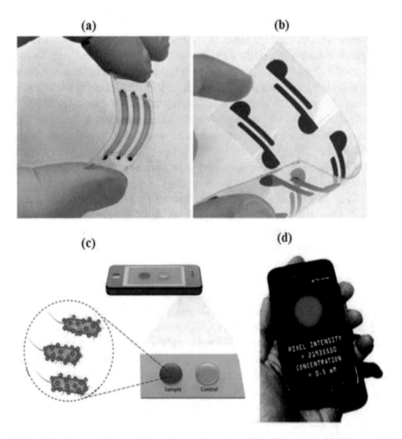

Fig. 3.14 µPAD device (**a, b**) used for pathogen detection with cell phone (**c, d**). (Sher et al. 2017)

Fig. 3.15 A μPAD for the detection of ARDS via neutrophil capture from patient blood. (Wu et al. 2018)

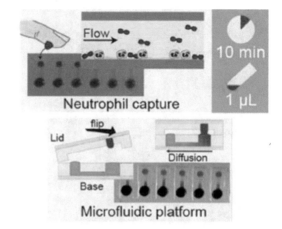

resulting color change in the nanoparticles can be easily detected by the naked eye, and a cell phone camera can be utilized to take a picture of the nanoparticle aggregation. But this is just the tip of the iceberg; the fact of the matter is that these modular devices hold immense promise for detecting an array of communicable and non-communicable diseases ranging from dengue fever to colon cancer (Lam et al. 2017). When it comes to humanitarian settings, one of the most critical conditions to treat are those known as acute respiratory distress syndromes (ARDS). This umbrella term consists of diseases such as pneumonia and is characterized by pulmonary inflammation as well as a mortality rate of 45% (Hansen and Lam 2017). The reason this condition is so deadly is due to its complex nature, which makes it difficult to identify and treat, ultimately resulting in diffuse alveolar damage and pulmonary microvascular endothelial injury leading to death (Hansen and Lam 2017). The pathogenesis of ARDS is primarily mediated by neutrophils; thus they make for ideal biomarkers in the diagnosis and detection of ARDS in patients before it is too late. Once again, μPADs make for an excellent diagnostic tool for this application as shown in Fig. 3.15. The μPAD depicted in this figure can efficiently and effectively detect neutrophil levels in a patient's blood for the diagnosis of ARDS in the field such as pneumonia.

μPADs are excellent platform for innovation as advances in paper and flexible material-based biosensing platforms make for powerful POC assays in resource-limited settings. The key to this continued innovation lies in the integration and incorporation with different detection strategies and technologies to enhance the detection and biosensing of an array of biological targets (Zarei 2017). For example, the integration of cellulose paper and flexible polyester films with optical biosensing platforms using antibodies and peptides has led to the enhanced detection of multiple biological targets including HIV, *Escherichia coli*, and *Staphylococcus aureus* and CD41 T lymphocytes as shown in Fig. 3.16 (Zarei 2017). This can all be done with just a single fingerprick's worth of blood and deliver clinical-level detection and sensitivity.

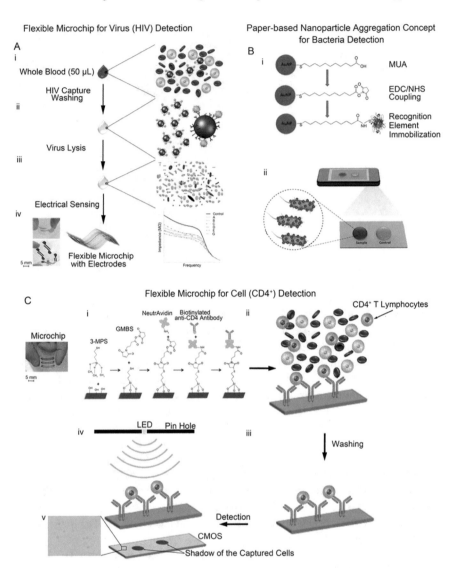

Fig. 3.16 Integrated paper and flexible substrate-based µPAD platforms. (**a**) HIV-detection via microchips on polyester film. (**b**) Paper-based nanoparticle aggregation for bacterial detection on a cellulose paper with smartphone camera. (**c**) CD41 T lymphocyte detection on polyester film. (Zarei 2017)

While indeed these devices are simple and user-friendly, µPADs still require a trained healthcare provider to interpret the data they provide and to prescribe any necessary treatments (So and Ruiz-Esparza 2012). In crisis and conflict situations, this is where the community health worker would serve as a provider, but since this book is on the humanitarian innovation, we couple these devices with telemedicine. Telemedicine has the revolutionary capacity to connect highly trained healthcare workers in countries around the world to patients in remote settings (So and Ruiz-Esparza 2012). This of course is made possible by substantial improvements to mobile communication technologies such as cell phones. Diagnostic results from a patient could be securely sent via smartphone to a physician both locally and internationally for quick diagnostic feedback and suggested treatment regimen. For example, a community healthcare worker would test a patient using a μPAD, photograph the results with a camera phone, and transmit the image to a central laboratory (So and Ruiz-Esparza 2012). An expert would then analyze the image and respond to prescribe an appropriate treatment regimen to pursue.

Over the course of this unique chapter, we have explored the feasible and functional deployment of several medical interventions and medical device innovations that would be appropriate for use in humanitarian settings. Once again, the functional deployment of these interventions lies in the bilateral transfer of knowledge and the cultivation of a suitable innovation environment that allows for the sufficient access to the building blocks of knowledge, the strategic use of intellectual property and innovative financing to meet public health goals, as well as the collaborative elements of multiple innovation paradigms including open, reverse, frugal, and disruptive innovation (So and Ruiz-Esparza 2012). In our next chapter, we further expand the realm of our inquiry into redefining innovation in humanitarian medicine and take an integrative approach beyond medical devices. In this next chapter, we hope to show how indeed innovation is cultivated in both a micro and macroscale that truly knows no bounds.

References

Behind the Fences of Jordan's Zaatari Refugee Camp. 2016. *Alarabiya.net*. Accessed 20 July 2018. https://english.alarabiya.net/en/blog/2016/03/28/Behind-the-fences-of-Jordan-s-Zaatari-refugee-camp.html

Bhatia, S. K., & Ramadurai, K. W. (2017). 3-Dimensional device fabrication: A bio-based materials approach. In *3D printing and bio-based materials in global health* (pp. 39–61). Cham: Springer.

Black, R., Laxminarayan, R., Temmerman, M., & Walker, N. (Eds.). (2016). *Disease control priorities, (Volume 2): Reproductive, maternal, newborn, and child health*. Washington, D.C: The World Bank.

Cotton, M., Henry, J. A., & Hasek, L. (2014). Value innovation: An important aspect of global surgical care. *Globalization and Health, 10*(1), 1.

Darzi, A. (2017). The cheap innovations the NHS could take from Sub-Saharan Africa. *The Guardian*. Accessed 25 July 2018. https://www.theguardian.com/healthcare-network/2017/oct/27/cheap-innovations-nhs-take-sub-saharan-africa

Demand generation I-Kit for underutilized, life saving commodities. 2018. *sbccimplementationkits.org*. Accessed 12 Aug 2018. https://sbccimplementationkits.org/demandrmnch/about-chx/

Drilling to the problem: Low-cost sterile drill covers for surgery. 2014. *Medgadget*. Accessed 24 July 2018. https://www.medgadget.com/2014/09/drilling-to-the-problem-low-cost-sterile-drill-covers-for-surgery-interview-with-lawrence-buchan-cofounder-of-arbutus-medical.html

Gathwala, G., Sharma, D., & Bhakhri, B. k. (2013). Effect of topical application of chlorhexidine for umbilical cord care in comparison with conventional dry cord care on the risk of neonatal sepsis: A randomized controlled trial. *Journal of Tropical Pediatrics, 59*(3), 209–213.

Gilmore, B., Adams, B. J., Bartoloni, A., Alhaydar, B., McAuliffe, E., Raven, J., Taegtmeyer, M., & Vallières, F. (2016). Improving the performance of community health workers in humanitarian emergencies: A realist evaluation protocol for the PIECES programme. *BMJ Open, 6*(8), e011753.

Hansen, C. E., & Lam, W. A. (2017). Clinical implications of single-cell microfluidic devices for hematological disorders. *Analytical Chemistry, 89*(22), 11881–11892.

Jabbar, S. A., & Zaza, H. I. (2014). Impact of conflict in Syria on Syrian children at the Zaatari refugee camp in Jordan. *Early Child Development and Care, 184*(9–10), 1507–1530.

Jonas, S. M., Deserno, T. M., Buhimschi, C. S., Makin, J., Choma, M. A., & Buhimschi, I. A. (2015). Smartphone-based diagnostic for preeclampsia: An mHealth solution for administering the Congo Red Dot (CRD) test in settings with limited resources. *Journal of the American Medical Informatics Association, 23*(1), 166–173.

Lam, T., Devadhasan, J. P., Howse, R., & Kim, J. (2017). A chemically patterned microfluidic paper-based analytical device (C-µPAD) for point-of-care diagnostics. *Scientific Reports, 7*(1), 1188.

Levin, D. (2015). Openideo – How might we improve education and expand learning opportunities for refugees around the world? – The World's first Fab Lab in a refugee camp. *challenges.openideo.com*. Accessed 20 July 2018. https://challenges.openideo.com/challenge/refugee-education/ideas/the-world-s-first-fab-lab-in-a-refugee-camp

Löfgren, J., Nordin, P., Ibingira, C., Matovu, A., Galiwango, E., & Wladis, A. (2016). A randomized trial of low-cost mesh in groin hernia repair. *New England Journal of Medicine, 374*(2), 146–153.

Makin, J., Suarez-Rebling, D. I., Varma Shivkumar, P., Tarimo, V., & Burke, T. F. (2018). Innovative uses of condom uterine balloon tamponade for postpartum hemorrhage in India and Tanzania. *Case Reports in Obstetrics and Gynecology, 2018*, 1.

Martinez, A. W., Phillips, S. T., Whitesides, G. M., & Carrilho, E. (2009). Diagnostics for the developing world: Microfluidic paper-based analytical devices. *Analytical Chemistry, 82*, 3–10.

McNeil, D. (2017). Car mechanic dreams up a tool to ease births. *nytimes.com*. Accessed 11 Aug 2018. https://www.nytimes.com/2013/11/14/health/new-tool-to-ease-difficult-births-a-plastic-bag.html

Namakula, J., & Witter, S. (2014). Living through conflict and post-conflict: Experiences of health workers in northern Uganda and lessons for people-centred health systems. *Health Policy and Planning, 29*(suppl_2), ii6–ii14.

Operation Hernia. 2018. *Operationhernia.org.uk*. Accessed 23 July 2018. http://operationhernia.org.uk/about-us/international-connections/

Scaling up life saving innovations for mothers and newborns. 2018. *options.co.uk*. Accessed 11 Aug 2018. https://www.options.co.uk/news/scaling-life-saving-innovations-mothers-and-newborns

Schvartzman, J. A., Krupitzki, H., Merialdi, M., Betrán, A. P., Requejo, J., Nguyen, M. H., Vayena, E., et al. (2018). Odon device for instrumental vaginal deliveries: Results of a medical device pilot clinical study. *Reproductive Health, 15*(1), 45.

Segre, J. 2013. Chlorhexidine for umbilical cord care: A best buy for newborn health. *Healthy Newborn Network*. Accessed 15 Aug 2018. https://www.healthynewbornnetwork.org/blog/chlorhexidine-for-umbilical-cord-care-a-best-buy-for-newborn-health/

Sher, D. (2015). Amazing project brings prostheses to Syrian refugee camp with help from ultimaker. *3Dprintingindustry.com*. Accessed 26 July 2018. https://3dprintingindustry.com/news/amazing-project-brings-prostheses-to-syrian-refugee-camp-with-help-from-ultimaker-44213/

Sher, M., Zhuang, R., Demirci, U., & Asghar, W. (2017). Paper-based analytical devices for clinical diagnosis: Recent advances in the fabrication techniques and sensing mechanisms. *Expert Review of Molecular Diagnostics, 17*(4), 351–366.

So, A. D., & Ruiz-Esparza, Q. (2012). Technology innovation for infectious diseases in the developing world. *Infectious Diseases of Poverty, 1*(1), 2.

Syedmoradi, L., & Gomez, F. A. (2017). Paper-based point-of-care testing in disease diagnostics. *Bioanalysis, 9*, 841–843.

UNICEF Innovation. 2013. Backpack PLUS toolkit created to help empower community health workers – Stories of innovation. *Stories of Innovation.* Accessed 19 July 2018. https://blogs.unicef.org/innovation/backpack-plus-toolkit-created-to-help-empower-community-health-workers/

Van Berlaer, G., Elsafti, A. M., Al Safadi, M., Souhil Saeed, S., Buyl, R., Debacker, M., Redwan, A., & Hubloue, I. (2017). Diagnoses, infections and injuries in Northern Syrian children during the civil war: A cross-sectional study. *PLoS One, 12*(9), e0182770.

Wu, J., Dong, M., Rigatto, C., Liu, Y., & Lin, F. (2018). Lab-on-chip technology for chronic disease diagnosis. *npj Digital Medicine, 1*(1), 7.

Zarei, M. (2017). Portable biosensing devices for point-of-care diagnostics: Recent developments and applications. *TrAC Trends in Analytical Chemistry, 91*, 26–41.

Chapter 4
Disruptive Technologies and Innovations in Humanitarian Aid and Disaster Relief: An Integrative Approach

In our previous chapters, we defined the nature of an array of innovation processes as well as their applications and development for humanitarian medicine. But once again, we have merely scratched the surface when it comes to the boundless potential for innovation in humanitarian response and management. In particular, we are currently in the midst of huge tangential shift in disaster response and mitigation responses employed by multilateral agencies. This is of course thanks to the rapid advancement and integration of novel technologies and user interfaces that hold promise in changing how we as humans respond to conflict and disaster. In this chapter, we take a holistic approach in examining a spectrum of disruptive technologies and innovation that seek to propel a new era of humanitarian medicine and the alleviation of human suffering. These innovations encompass an array of fields ranging from data collection and crisis management to the integration of mobile health (mHealth) applications and blockchain platforms to better treat refugees and conflict victims. What is fascinating is that each one of these innovations has the innate capacity to revolutionize the humanitarian relief paradigm. But when these innovations are put together, they create a highly dynamic, functional innovation ecosystem that serves as an impetus for unlimited value-creation and problem-solution discovery.

4.1 Data Collection and Crisis Management: Crowdsourced Crisis Mapping

When dealing with humanitarian crises, the logistical component is just as important as the functional intervention. Specifically, the strategic management of resources and human capital in conjunction with data collection is vital in crafting a quick and efficient response. But how do we sufficiently gauge the functional deployment

© The Author(s), under exclusive licence to Springer Nature Switzerland AG 2019   75
K. W. Ramadurai, S. K. Bhatia, *Reimagining Innovation in Humanitarian Medicine*,
SpringerBriefs in Bioengineering, https://doi.org/10.1007/978-3-030-03285-2_4

resources among the chaos and confusion present in many situations? This is where crisis mapping comes into play. Of course, crisis and GIS (geographic information system) mapping is not a novel concept but has recently found innovation in its applications beyond just mapping. GIS is a way of visually presenting, analyzing, and managing data and statistics, primarily through the use of maps (Gao et al. 2011). Crisis mapping is a spin-off of GIS, which utilizes a larger set of tasks that not only provide a geolocated visualization but also for the filtering, categorization, and analysis of information (Norheim-Hagtun and Meier 2010). These mapping platforms have been uniquely adapted for data analysis and management as well as a visualization tool. For example, GIS mapping was extremely useful in the recent Ebola outbreak, in which it was utilized to digitally map and visualize outbreak locations and casualties (Gao et al. 2011). Furthermore, GIS mapping was deployed in crisis mapping during the Haitian earthquake in order to sufficiently gauge aid supply need (Norheim-Hagtun and Meier 2010). As with all technological innovations, the technology represents only one side of the problem. The other side is our ability to bridge the gap between the creation and sharing of knowledge and creating effective interventions with that knowledge (Gao et al. 2011). The innovation we explore is that of "crowdsourced crisis mapping," which aims to harness the streams of information that flow through social media to provide response organizations with near-real-time, categorized, and geolocated data (Gao et al. 2011). The explosion of user-generated content through social media can be leveraged to assist first-aid responders and humanitarian organizations in the wake of natural disasters, crises, and violent conflicts (Norheim-Hagtun and Meier 2010).

As with any humanitarian intervention, coordination is a central, challenging issue in agent-oriented distributed systems (Heinzelman and Waters 2010). Disaster relief systems and agencies primarily focus on designing coordination protocols and mechanisms to manage governmental and nongovernmental organization activities. Research has shown that it is possible to leverage social media to generate community crisis maps and introduce an interagency map to allow organizations to share information as well as collaborate, plan, and execute shared missions (Fig. 4.1) (Gao et al. 2011). Crowdsourcing allows individuals to participate in various tasks ranging from information validation to data management related to information-sharing communities. The information garnered from these social media platforms

Fig. 4.1 Crowdsourcing interagency map between public and relief organizations. (Gao et al. 2011)

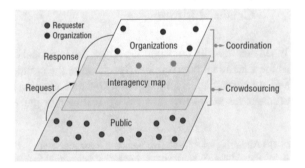

can then be shared among public and private agencies in order to create effective interventions (Fig. 4.1).

The rapid growth of crowdsourcing applications for disaster relief and humanitarian medicine is directly benefited from the data acquired from various sources in a short time. Organizations such as Ushahidi, utilize an open source crisis map platform that has been deployed in Kenya, Mexico, Afghanistan, and Haiti. This platform leverages web technologies to integrate data from multiple sources including mobile phones, email, and social media sites such as Twitter and Facebook to provide an up-to-date, publicly available crisis map that is in turn available to relief organizations (Meier 2012). This platform uses crowdsourcing for social activism and public accountability to collectively contribute information, visualize incidents, and enable cooperation among various organizations (Meier 2012). When it comes to humanitarian medicine, the implications of crowdsourced information platforms are quite remarkable. These platforms can be utilized to identify infectious disease hotspots from local sources and serve as a dynamic interface for medical interventions in field. For example, the integration of mobile telehealth with crowdsourced information means that healthcare provides on the ground could pose unconventional medical inquiries and situations to global network interface for direct feedback.

With regard to the humanitarian response as a whole, crowdsourced data including user requests and status reports can be collected almost immediately after a disaster using social media. Ushahidi Haiti was set up 2 h after the January 12, 2010, earthquake by volunteers from Tufts University (Meier 2012). Soon after, organizations were able to utilize a free SMS emergency number that was spread via national radio. By January 25th of 2010, the Haiti crisis map had more than 2500 incident reports, in which the large amount of nearly real-time reports allowed inter-agency cooperation and the ability to identify and respond to urgent cases in time (Meier 2012). Crowdsourcing tools often utilize multiple data inputs from emails, forms, and even social media "tweets," which can be meta-analyzed to create forecast trends and segregate data by urgency (Heinzelman and Waters 2010). The categorical classification allows for the streamlined organization and determination of medical, food, and shelter needs for the disaster or conflict-stricken victims. For example, Fig. 4.2 illustrates the food requests via the Ushahidi Haiti interface map which allowed organizations to coordinate resource distribution from crowdsourced data on the ground. The innovation of is further extended into the "geo-tagging" of information for messages sent from social media platforms such as Twitter as well as cellular devices (Heinzelman and Waters 2010). Geo-tagging allows for the creation of live-time crisis maps in war-torn areas such as Syria shown in Fig. 4.3.

While indeed the innovation harbored by crowdsourced crisis mapping is quite novel with true implications for revolutionizing the way medical care and aid delivery is allocated in unconventional environments, there is indeed room for improvement. Current crisis mapping applications do not provide a common mechanism specifically designed for coordination among relief organizations. For example, crisis maps do not provide a mechanism for apportioning response resources; thus multiple organizations might respond to an individual request at the same time (Gao et al. 2011). This of course can mean that there is an interventional "clash" among

Fig. 4.2 Ushahidi Haiti crisis map for the logistical support and allocation of food resources. (Crowdsourced Crisis Mapping 2012)

aid-granting organizations which may all respond to one incident, while bypassing others. Furthermore, there is a propensity for the creation of inaccurate geo-tags in determining the actual spatial features on the ground. It is also vital that crowd-sourced data be supplemented by a group-sourced relief situation summary, verified geotag information, and sufficiently relayed to teams on the ground. Report verification and validation is absolutely critical, as the creation of unverified claims, events, or unverified geo-tags could significantly hinder response time and resource allocation. Thus, further refinement and enhancement of the validation process in the crowdsourced data analysis is vital in further enhancing the interventional capacity of crisis mapping.

4.2   Robotics and Wearable Technology

Our next segment explores the feasible yet functional deployment of innovations related to robotics and wearable technologies in humanitarian medicine. Specifically, we examine two frugal and disruptive innovations related to "Lab-on-a-drone" systems and wearable technology for surgical teleproctoring (Zarei 2017; Datta et al. 2015). The integration of these two innovations holds promise in enhancing the way medicine and clinical diagnosis is delivered in crisis and conflict situations. These innovations are intriguing as they require little with regard to infrastructure and represent a shift in the deployment of affordable and feasible technologies in solving humanity's most pressing problems. Furthermore, they can be scaled and developed in a variety of contexts, environments, and situations. The true innovation in these technologies not only lies in their interventional capacities, but also in their capacity for reverse innovation and global dissemination.

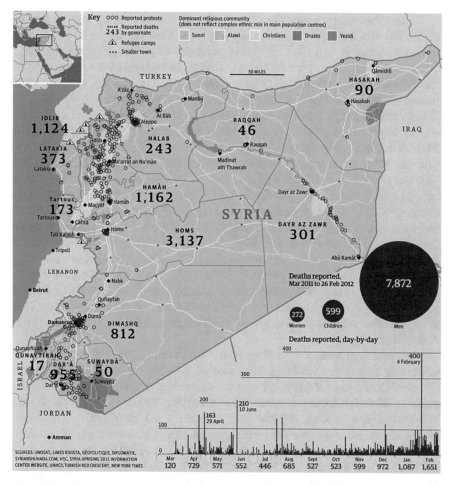

Fig. 4.3 Crowdsourced real-time geo-tagging of conflict deaths in Syria. (Crowdsourced Crisis Mapping 2012)

Unmanned aerial vehicles (UAVs)—commonly referred to as "drones"—offer humanitarian agencies truly endless applications ranging from crisis mapping to search and rescue as well as supply transport (Zarei 2017). The recent advent and development of drone technology has been vastly enhanced in only the past decade. With these rapid developments, drone technology has become more affordable and better-engineered—so much so that the United Nations Office for the Coordination of Humanitarian Affairs (UNOCHA) has issued new guidelines on the use of drones in humanitarian efforts and disaster response (Martini et al. 2016). But while indeed much previous literature has explored the development of drones for mobile surveillance and other more conventional means, in this section we explore a truly novel application known as "lab-on-a-drone."

Current laboratory testing for clinical diagnoses and existing approaches are highly resource intensive. These testing protocols often rely on the deployment of trained individuals to gather biological samples, which are then collected and returned to laboratories for analysis. Of course, in humanitarian scenarios, these channels are often inefficient as there is often a lack of transportation and medical infrastructure. This means that samples collected in a refugee or internally displaced victim camp experience considerable time lag between sample collection, diagnosis, and implementation of treatment interventions (Martini et al. 2016; Priye et al. 2016). This time lag in sample analysis could mean the difference between life and death for patients that are experiencing chronic illness or infectious disease. This is where the contextualized adaptation of drones for laboratory sample collection, transportation, and analysis comes into play, i.e., lab-on-a-drone. The lab-on-a-drone system is a highly mobile biochemical analysis platform for rapid field deployment of nucleic acid-based diagnostics utilizing basic, consumer-grade quadcopter drones (Priye et al. 2016). With regard to the formal operation of the lab component of the drone, the platform isothermally performs the polymerase chain reaction (PCR) with a single heater. This enables the system to be operated using standard USB power source in the form of a small battery—the same type used to power mobile devices. The fluorescence detection and quantification of the PCR sample is achieved utilizing a simple smartphone camera for image analysis (Priye et al. 2016).

With regard to standard sample preparation, the drone's actual motors can be utilized as centrifuges via 3D-printed snap-on attachments as shown in Fig. 4.4 (Zarei 2017; Priye et al. 2016). These advancements make it possible to build a complete DNA/RNA analysis system at a cost of only $50 USD (Priye et al. 2016). This makes for a low-cost, durable, highly versatile, and adaptable mobile laboratory platform that allows for the precise deployment of molecular diagnostic techniques in field sites in any area or environment. The efficacy behind this technology has been reflected in multiple field tests which has displayed successful in-flight replication of *Staphylococcus aureus* and λ-phage DNA targets in rapid time (Zarei 2017; Martini et al. 2016; Priye et al. 2016). The ability to perform rapid in-flight assays with smartphone connectivity is an extremely novel innovation. This closes the gap in time between sample collection and analysis; thus test results can be delivered in rapid time. The scope of applications for this technology is quite profound. Lab-on-a-drone could be utilized during infectious disease epidemics where the ability to rapidly diagnose patients in rural areas is vital such as during the Ebola epidemic or cholera outbreaks experienced in sub-Saharan Africa and Haiti, respectively. Furthermore, the applications for humanitarian medicine are quite profound, as lab-on-a-drone could allow disease diagnosis in extremely unconventional environments such as war zones or refugee camps, where suitable healthcare infrastructure is not present. The contextual adaptation drone-based technology in conjunction with other innovations such as 3D printing represents a new era in combinatorial innovations that extend beyond the convention roles of drones in supply transport and surveillance video/imaging. Once again, the feasible deployment of these technologies lies in the culmination of the humanitarian innovation ecosystem.

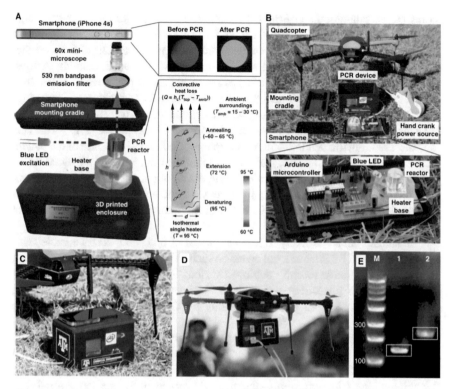

Fig. 4.4 Lab-on-a-drone system. (**a**) Isothermal PCR using a single battery-powered heating apparatus. (**b**) Components of the lab-on-a-drone platform. (**c**) Smartphone camera integration for fluorescence detection. (**d**) Attachment of the laboratory enclosure on a basic, consumer-grade quadcopter drone. (**e**) Successful in-flight PCR of two different DNA targets. (Priye et al. 2016)

This means that the future development of these technologies for unconventional applications must be developed via a crowdsourced network platform both on the ground as well as digitally. Creating a direct interface between deployment on the ground and researchers abroad via crowdsourced technology platforms such as the internet allows the rapid advancement of the technology for field applications. Furthermore, fostering the innovation ecosystem on the ground allows for practitioners, aid workers, and conflict victims/refugees to have first-hand input into how the device can further be modified and enhanced.

The next innovation we explore related to humanitarian medicine is that of wearable technologies. These innovations refer to technologies that can be worn on our person in order to provide direct and instantaneous feedback related to our health and associated conditions. These technologies range in scope and application, but all hold immense promise in revolutionizing the way healthcare is delivered in unconventional, resource-poor settings. The first innovation we explore is called "Khushi Baby," which means *happy baby* in Hindi, and is a basic plastic pendant on a black string (Khushi Baby Case Study by UNESCO-Pearson Initiative for Literacy 2017). Contained within this simple plastic pendant is a small computer chip which stores the vaccination data of the baby wearing it, along with the mother's health records (Fig. 4.5) (Khushi Baby 2018). Having this information contained within a small, durable device such as this pendant holds immense potential value for on-the-move populations such as refugees and internally displaced conflict victims. Keeping the health records and information of babies in this device, rather than medical cards/folders, allows health workers the capacity to make sure that babies receive proper vaccination treatment. The data collection process from the Khushi pendant is user-friendly and streamlined into three easy steps (Fig. 4.6). This includes scanning the pendant with a smartphone, updating the profile of the patient on the smartphone application, and then syncing the data to Khushi's cloud-based platform. This allows for the instantaneous updating of patient information and vaccination protocols for babies. This simple device could mean the difference between life and death for infants of whom are part of migratory populations of refugees in conflict situations.

The Khushi device is likely to join an arsenal of other low-cost wearable technologies garnered by international agencies such as UNICEF. In fact, organizations such as UNICEF have recently deployed various low-tech wearables, such as the multicolored arm measuring tape, which displays whether a child is receiving enough nutrition (Khushi Baby Case Study by UNESCO-Pearson Initiative for Literacy 2017).

The next wearable technology innovation we delve into focuses on acute trauma care. Oftentimes in emergency situations, the ability to measure and monitor vital signs of trauma patients at the point of injury is a critical facet of care. Unfortunately, this still remains a time-consuming and manual process for first responders and doctors that are operating in the field. In response to this problem, the First Response Monitor was created by the Cambridge Design Partnership. The First Response Monitor enables rapid, reliable measurement and real-time display of a patient's respiratory and heart rate. The device further records data to allow access to vital signs trend data over time and is connected to a smartphone or tablet for data analysis of multiple patients in triage (First Response Monitor | Cambridge Design Case Studies, 2018). The device, as shown in Fig. 4.7, is simply clipped onto the nasal passage of a patient and immediately begins to track vital signs. This data is then

Fig. 4.5 Khushi Baby mobile device technology and microchip apparatus. (Khushi Baby's Necklace Keeps Track of Immunizations 2016)

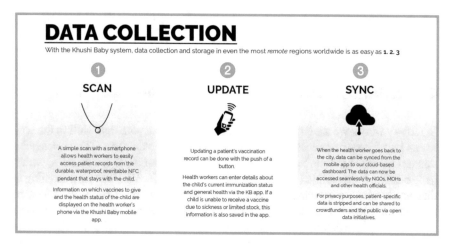

DATA COLLECTION

With the Khushi Baby system, data collection and storage in even the most *remote* regions worldwide is as easy as **1. 2. 3**

① SCAN

A simple scan with a smartphone allows health workers to easily access patient records from the durable, waterproof, rewritable NFC pendant that stays with the child.

Information on which vaccines to give and the health status of the child are displayed on the health worker's phone via the Khushi Baby mobile app.

② UPDATE

Updating a patient's vaccination record can be done with the push of a button.

Health workers can enter details about the child's current immunization status and general health via the KB app. If a child is unable to receive a vaccine due to sickness or limited stock, this information is also saved in the app.

③ SYNC

When the health worker goes back to the city, data can be synced from the mobile app to our cloud-based dashboard. The data can now be accessed seamlessly by NGOs, MOHs and other health officials.

For privacy purposes, patient-specific data is stripped and can be shared to crowdfunders and the public via open data initiatives.

Fig. 4.6   Khushi Baby data collection process. (Khushi Baby 2018)

Fig. 4.7   First Response Monitor device. (First Response Monitor by Cambridge Design Partnership 2015)

compiled in a trend graph that assesses real-time changes in a patient's condition and vital signs. The First Response Monitor is a low cost, durable, high-utility device that mitigates the need to complex monitors and equipment when operating in low-resource settings. With sufficient economies of scale, intuitive devices such as this one have immense potential in being easily disseminated in aid/relief supply kits in field-operating clinics.

In looking at the future, the integration of technology into a holistic interface will prove to be the impetus in fostering innovation in humanitarian medicine. The true interventional capacity of these technologies to save lives around the world is exponentially increased when utilized in conjunction with one another. The culmination of technologies such as the ones mentioned in this section seeks to vastly enhance the relative abilities and capacities of medical practitioners to treat patients. In our next section, we explore the advent of mHealth, telemedicine, and blockchain, which are dynamic platforms and interfaces that serve to radically change the way medicine is delivered in the digital age.

4.3   mHealth, Telemedicine, and Blockchain

In today's world, we are currently experiencing a massive paradigm shift in the way we as humans communicate and exchange information and data. The impetus for this lies in that of mobile technologies such as most prominently cellphones, which have become a streamlined technology in only the last decade. What is intriguing is that the barriers to dissemination of mobile technology have been quite low, in which people in both developed and developing countries have ubiquitous access to cellphones. It is estimated that there are more than 7 billion mobile or cellular network subscribers globally, with the majority being in low- and middle-income countries (Bastawrous and Armstrong 2013). When it comes to refugee and internally displaced victims, cellphones are less prevalent than the global population, but yet approximately 80% of refugees have access to 2G and 3G network connections (Bastawrous and Armstrong 2013). With the advent and rapid dissemination of mobile technologies, the field of humanitarian medicine has the distinct opportunity and ability to vastly enhance patient data collection, improve quality of care, as well as connect healthcare workers and aid organizations. Furthermore, mobile technology provides a powerful tool that can be utilized by crisis-afflicted populations to make their own healthcare decisions (Doocy et al. 2017).

Organizations such as the UNHCR have directly recognized the vast potential for mobile platforms to improve the lives of refugees and host communities. The output of this realization is the development of the Global Strategy for Connectivity for Refugees Mobile technologies to develop eHealth (electronic health) and mHealth (mobile health) infrastructure to support frontline health workers (Doocy et al. 2017; Mesmar et al. 2016). This would allow access to patient information and data in the field for practitioners to utilize instantaneously for treatment assessment and diagnoses. This is a vital tool for community health workers which often serve as the face of primary care delivery in low-resource settings. In looking closer at mHealth and eHealth, the true essence of its novelty must be embraced in functional deployment in the field. Recently, there have been an array of studies that have examined the utility value and feasibility of utilizing these technologies in the field. For example, in Lebanon, a mHealth app was developed in order to improve the quality of care for refugee patients with hypertension and/or type 2 diabetes at primary healthcare centers. The mHealth app known as "Sana mobile" was based on a tablet device, allowing physicians and patients to keep track of their medical records (Doocy et al. 2017). More than 4.5 million Syrians have fled to neighboring countries including Lebanon, which is host to the highest number of refugees per capita in the world (Doocy et al. 2017). This high displacement of refugees has taken an immense toll on Lebanon's healthcare infrastructure, making it difficult to sufficiently treat and discharge patients with chronic diseases such as diabetes and hypertension. Thus, the advent of mHealth technologies seeks to streamline refugee patient data collection into a mobile platform that can be utilized in unconventional environments.

The estimated prevalence of type 2 diabetes is more than 7% in Syria and 14% in Lebanon, with a prevalence of hypertension greater than 25%, in Syria and Lebanon (Doocy et al. 2017; Mesmar et al. 2016). Given the high burden of noncommunicable diseases (NCDs) among refugees and their host country populations, humanitarian agencies and practitioners face immense challenges in providing sufficient medical care and treatment needs. These NCD conditions are further difficult to manage as not only are they are chronic in nature, but the resources for refugee care are limited. Utilizing mobile devices can indeed bridge this gap in care and improve the overall public health of a population across an array of disciplines and applications. These range from patient education, point-of-care diagnostics, electronic health records, patient-provider communication, as well as healthcare provider training and education (Mesmar et al. 2016). The use of the Sana mobile app allowed enabled physicians to provide refugee patients with their disease history and medications prescribed at every visit by sending an SMS directly to the patient's phone (Perakslis 2018). The application could also send and deliver appointment notifications and reminders to the patient's phone to enhance treatment adherence. The application enhances the patient's ability to monitor their own health condition, as well as provides easy access to medical information in any setting.

Of course, with the advent of any novel technology, there will be issues related to adoption and functional implementation of the technology in everyday use. For mHealth technologies such as Sana, this was just the case to an extent. While indeed the overall reporting of clinic medical records and application uptake remained low, during the mHealth phase, recording of BMI and blood pressure was significantly greater in the mHealth application as compared to clinic medical records (Doocy et al. 2017). Patients reported more frequent measurements of weight, height, blood pressure, and blood glucose and were more satisfied with their clinic visits. This suggests that the mHealth application enhanced the physician's ability to record patient data, resulting in enhanced patient care dynamics. The outcomes of interventional studies such as this serve as a baseline in strategically developing and enhancing the capacities of these technologies to improve patient care. The culmination of these studies yields data that can outline the challenges and best strategies to improve eHealth and mHealth delivery as shown in Table 4.1.

mHealth technologies such as Sana and Khushi Baby represent disruptive innovations in patient record-keeping and allow for the secure and reliable access to vital patient data and information for conflict victims and refugees. But once again, these technological innovations represent only one piece of the puzzle. While indeed they are low-cost and interventionally feasible, the reality is that they must be integrated into a mHealth and eHealth digital platform. The integration of auxiliary mHealth technologies, i.e., mobile applications, into a digital platform/interface allows for medical practitioners to access patient data and reports from any location. The next innovation frontier related to this is the integration of blockchain into a comprehensive network interface to not only streamline patient data and record-keeping, but the

Table 4.1 Best practices and challenges faced in eHealth and mHealth implementation. (Perakslis 2018)

Challenges in eHealth and mHealth Delivery	Best practices and specific methodologies
Poor or limited user involvement and engagement	User-centered design, user co-design, and participatory design methodologies
Unclear goals, expectations, and scope creep	Develop and use a clear requirements Si expectations matrix
Poor sponsor participation and active leadership	Document-specific sponsor role requirements and the corresponding relationships to other roles
Poor technology selection	Use an established technology selection framework
Lack of necessary technology skill sets	Understand the necessary roles and recruit, train, or contract
Poor project management and lack of formal methodology	Understand and select from six most common technology delivery methodologies

humanitarian ecosystem as a whole. This innovative platform allows for the integration and application of novel technologies such as telehealth, feasible in disaster and conflict areas around the world. In our next segment, we explore the exciting frontiers and implications of telemedicine and blockchain to alleviate human suffering.

Telehealth is a digital innovation that offers immense promise in enhancing global access to reliable healthcare. Telehealth involves the remote exchange of medical information via medical professional to improve patient care, to educate patients, or to educate healthcare providers (Staruch et al. 2018; Pupic 2017). Telehealth has limitless potential in all spectrums of the medical field ranging from primary care to global surgery. Telehealth is indeed different from mHealth, as it is more focused on clinical care, rather than solely capturing data on patients (Staruch et al. 2018; Pupic 2017). Telehealth involves clinician-directed remote patient monitoring technology and encompasses clinician-to-clinician and clinician-to-patient interaction. This interaction has recently been extended to new frontiers including teleproctoring, which has been deemed as an extremely effective means of remotely sharing medical expertise in the field. Through this process, physicians and medical specialists can provide educational access as well as competency-based training to practitioners in a remote location (Agnisarman et al. 2017). The benefits of this capability are exponentially greater in resource-poor settings with otherwise limited access to human capital, peer mentorship, and reliable evaluation of medical training and techniques. Telehealth comprises of interactive medicine, i.e., patient interaction, the HIPPA-compliant sharing of patient information among practitioners, and remote diagnostic patient monitoring of vitals and other bodily functions (Agnisarman et al. 2017; Ekanoye et al. 2017; Krishnan et al. 2016; Walji 2015). There are indeed clear benefits to the implementation of telehealth initiatives, but once again like any technology, there are barriers to entry. Since telehealth is a relatively new and novel technology, it has yet to achieve sufficient economies of scale in order to be fully cost-effective and affordable for use in humanitarian applications. Telehealth involves not only mobile equipment including a tablet, microphone, screen, a proper network, and additional digital communications equipment, but also it requires the input of medical providers to make it work properly. The amalgam of these requirements represents a challenge to the full deployment of telehealth technologies in the field, but nonetheless there is still much promise. For example, one telehealth platform innovation that has been functionally deployed in humanitarian crisis and conflict situations is called "VSee." VSee is a proprietary low-bandwidth, group video chat and screen-sharing software tool and platform that was developed in order to tackle the problem of making virtual teamwork easy over video (Agnisarman et al. 2017; Ekanoye et al. 2017; Walji 2015). The service allows multiple users in various locations to communicate in real-time via video and audio over a basic 2G or 3G cellular network. The system has been deployed in humanitarian medicine operations around the world including Iraq, Kurdistan, Syria, and Nigeria (Fig. 4.8) (Agnisarman et al. 2017; Ekanoye et al. 2017; Krishnan et al. 2016).

We briefly mentioned previously that innovations such as telehealth are integrative in nature and present novelty when utilized in conjunction with other technological platforms. With regard to technological innovations such as VSee, we explore a new innovative interface that can be utilized in complement to further enhance humanitarian medicine and aid delivery in the modern era—this being blockchain. The term "blockchain" has surged in popularity in only the past 2 years, thanks in part to the dawn of a new era of currency known as "cryptocurrency." The most famous example of cryptocurrency is Bitcoin, which was the world's first unregulated free market currency. But for the sake of this book, we do not explore the innovation aspect of cryptocurrency, but rather the behind-the-scenes analytics platform known as blockchain. When it comes to humanitarian medicine, the ability to provide health services in an efficient and effective manner to conflict and refugee populations is paramount. Due to limited financial resources allocated toward modern refugee crises in around the world, the need for a platform that can streamline logistical support and efficiently delegate and harbor resources related to humanitarian operations in a sustainable manner has become critical. Blockchain was originally created to transfer financial value but has now become noted as an efficient and secure way to share information. But before we delve further into blockchain, there are three critical elements to realize (Fig. 4.9) (Blockchain for the Humanitarian Sector 2017):

Fig. 4.8 Use of the VSee telehealth mobile apparatus for remote consultation of ophthalmologic condition in patient in the Syrian refugee camp of Domiz. (VSee Team in Iraq Kurdistan Refugee Telemedicine Clinic 2014)

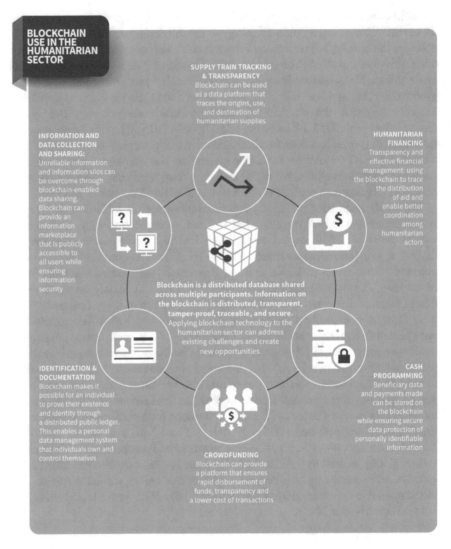

Fig. 4.9 Integration of blockchain into humanitarian logistics. (Blockchain for the Humanitarian Sector 2017)

1. Blockchain technology can be applied to humanitarian challenges, but it is not a separate humanitarian innovation in itself.
2. Blockchain's full potential can be realized in it use and deployment with other innovations such as Telehealth, eHealth, and mHealth.
3. In the humanitarian sector, some potential use cases for blockchain technology include information management, supply chain tracking, and humanitarian financing.

In crisis and conflict situations, oftentimes refugees, healthcare, and education are simply pushed to the side due to the immediate relief need. Given the fact that many refugees and conflict victims have chronic conditions, the need for adequate care and patient data collection is vital. Many times, patient data is recorded on simple Excel spreadsheets that are inaccurate due to incorrect name spelling, incorrect/unknown birth dates, as well as the input of multiple entries for the same patient. Currently, refugees carry papers containing doctor notes in addition to supplemental materials such as X-rays or lab-work (How Tech Can Bring Dignity to Refugees in Humanitarian Crises 2018). These materials are very rudimentary and do not offer much with regard to current treatment or diagnosis, since there is no access to a reliable patient medical history. Thus, having comprehensive medical records is critical, especially when it comes to immunization schedules when populations and refugees migrate through multiple countries (How Tech Can Bring Dignity to Refugees in Humanitarian Crises 2018). It is clear that there is much need to improve patient medical records, which has opened the door for cross-collaborations among telehealth and blockchain companies. For example, the Dubai-based telemedicine company Ver2 and the Slovenian healthcare blockchain company Iryo have recently collaborated to create the world's first blockchain-based telehealth platform (An Insight into the Life-Changing Aspects of Blockchain 2018). The inspiration for this collaboration came when Ver2 Founder Brian de Francesca joined and then took a trip to a Syrian refugee camp in the Bekka Valley, Lebanon. His purpose was to support doctors by providing administrative functions in a mobile clinical care setting. While he was there, he had several startling revelations, as he states that "One young Syrian girl had a large rash on her face that the two doctors couldn't diagnose. It had been there for many months. I took a picture of the rash with my smartphone and sent it to a German dermatologist working in Dubai. Two hours later, the doctor replied with the diagnosis and treatment plan—which was nothing more than a simple cream that could be purchased in the town nearby" (An Insight into the Life-Changing Aspects of Blockchain 2018).

It was during this encounter that Francesca realized the benefit of basic telehealth technologies for refugee and conflict victim care. But there was a caveat to this notion, as obtaining secure, accurate, and authenticated medical records would be vital in deploying telehealth technologies. This is where the innovation was derived, as blockchain could serve as the perfect platform that could enable the storage and access of these patient records. Thus, the world's first cloud-based blockchain-enabled patient-owned record system was born and developed. Typically, patients in refugee camps share one smartphone per family, with a 3G-enabled data connection, which can store health data on their mobile phones. This means that refugee families can have a mobile medical history while migrating away from conflict areas. This portable medical history would enhance continuous care with patient-centric, data-driven decision-making.

In order to functionally apply Iryo's blockchain technology in the real world, the company partnered with "Walk With Me," a US-based nonprofit organization that provides community health services, vocational training, and education to conflict victims and refugees around the world (Zajc 2018). The organization currently operates 12 projects in Iraq, Syria, Jordan, Lebanon, Egypt, and Djibouti providing services to more than ten million refugees (Zajc 2018). The goal is to provide these project sites with basic IT infrastructure to improve the quality of healthcare for refugees both in the present and future. The true impact of the technology will be harbored in the longitudinal medical data that is collected during the timeframe that refugees traditionally spend in camps—an average of 5–7 years (Zajc 2018). The initial use of the Iryo system will begin in refugee camps located throughout the Beqaa Valley of Lebanon and will eventually be scaled to other countries. During the coming years, Ver2 will expand the blockchain telehealth platform to include teleconsultations, education, and physician support services (Zajc 2018).

Once again, we see how the true essence of innovation is derived not from a single entity, device, or technology, but rather the integration of these innovations that work together to create a novel, functional interface. The innovations that we have analyzed in the section represent the forefront of new frontiers of innovation in humanitarian medicine. What we see is that not only is the innovation process important but also the function engineering of these devices and technologies and ultimately the formal deployment of them to alleviate human suffering in the field. The humanitarian landscape is rapidly changing, but it is important that we keep in mind the recipients of these technologies and the true power they hold in their ability and capacity to innovate as well. We advocate for a reciprocal innovation process of forward and reverse innovation, whereby engineering knowledge is bidirectional from refugee/patient/conflict victim to the innovation purveyor and vice versa.

References

Agnisarman, S., Narasimha, S., Madathil, K. C., Welch, B., Brinda, F. N. U., Ashok, A., & McElligott, J. (2017). Toward a more usable home-based video telemedicine system: A heuristic evaluation of the clinician user interfaces of home-based video telemedicine systems. *JMIR Human Factors, 4*(2), e11.

An insight into the life-changing aspects of blockchain. 2018. *Medtechengine.com*. Accessed 29 July 2018. https://medtechengine.com/article/blockchain/

Bastawrous, A., & Armstrong, M. J. (2013). Mobile health use in low-and high-income countries: An overview of the peer-reviewed literature. *Journal of the Royal Society of Medicine, 106*(4), 130–142.

Blockchain for the Humanitarian sector – Future opportunities. 2017. *UN Blockchain*. Accessed 26 Aug 2018. https://un-blockchain.org/2017/05/03/blockchain-for-the-humanitarian-sector-future-opportunities/

Crowdsourced crisis mapping: How it works and why it matters. 2012. *The Conversation*. Accessed 21 Aug 2018. https://theconversation.com/crowdsourced-crisis-mapping-how-it-works-and-why-it-matters-7014

Datta, N., MacQueen, I. T., Schroeder, A. D., Wilson, J. J., Espinoza, J. C., Wagner, J. P., Filipi, C. J., & Chen, D. C. (2015). Wearable technology for global surgical teleproctoring. *Journal of surgical education, 72*(6), 1290–1295.

Doocy, S., Paik, K., Lyles, E., Tam, H. H., Fahed, Z., Winkler, E., Kontunen, K., Mkanna, A., & Burnham, G. (2017). Pilot testing and implementation of a mHealth tool for Non-communicable Diseases in a Humanitarian Setting. *PLoS currents, 9*, 5–42.

Ekanoye, F., Ayeni, F., Olokunde, T., Nina, V., Donalds, C., & Mbarika, V. (2017). Telemedicine diffusion in a developing country: A case of Nigeria. *Science Journal of Public Health, 5*(4), 341.

First response monitor | Cambridge design case studies. 2018. *Cambridge-Design.Com.* Accessed 27 Aug 2018. https://www.cambridge-design.com/case-studies/first-response-monitor

First response monitor by Cambridge design partnership. 2015. *InnovaSystems.* Accessed 23 Aug 2018. https://www.innova-systems.co.uk/first-response-monitor-cambridge-design-partnership/

Gao, H., Barbier, G., Goolsby, R., & Zeng, D. (2011). *Harnessing the crowdsourcing power of social media for disaster relief.* Tempe: Arizona State University.

Heinzelman, J., & Waters, C. (2010). *Crowdsourcing crisis information in disaster-affected Haiti.* Washington, DC: US Institute of Peace.

How tech can bring dignity to refugees in humanitarian crises. 2018. *The Conversation.* Accessed 20 July 2018. https://theconversation.com/how-tech-can-bring-dignity-to-refugees-in-humanitarian-crises-94213

Khushi Baby. 2018. *Khushibaby.org.* Accessed 25 Aug 2018. https://www.khushibaby.org/

Khushi Baby case study by UNESCO-Pearson Initiative for Literacy. *UNESCO,* 2017. http://unesdoc.unesco.org/images/0026/002605/260596E.pdf

Khushi Baby's necklace keeps track of immunizations. 2016. *Futurity.* Accessed 21 Aug 2018. https://www.futurity.org/immunizations-india-khushi-baby-1084412-2/

Krishnan, G., Chawdhry, V., Premanand, S., Sarma, A., Chandralekha, J., Kumar, K. Y., Kumar, S., & Guleri, R. (2016). Telemedicine in the Himalayas: Operational challenges—A preliminary report. *Telemedicine and e-Health, 22*(10), 821–835.

Martini, T., Lynch, M., Weaver, A., & van Vuuren, T. (2016). The humanitarian use of drones as an emerging technology for emerging needs. In *The future of drone use* (pp. 133–152). The Hague: TMC Asser Press.

Meier, P. (2012). Ushahidi as a liberation technology. In *Liberation technology: Social media and the struggle for democracy* (pp. 95–109). Baltimore, MD: Johns Hopkins Press.

Mesmar, S., Talhouk, R., Akik, C., Olivier, P., Elhajj, I. H., Elbassuoni, S., Armoush, S., et al. (2016). The impact of digital technology on health of populations affected by humanitarian crises: Recent innovations and current gaps. *Journal of Public Health Policy, 37*(2), 167–200.

Norheim-Hagtun, I., & Meier, P. (2010). Crowdsourcing for crisis mapping in Haiti. *Innovations: Technology, Governance, Globalization, 5*(4), 81–89.

Perakslis, E. D. (2018). Using digital health to enable ethical health research in conflict and other humanitarian settings. *Conflict and Health, 12*(1), 23.

Priye, A., Wong, S., Bi, Y., Carpio, M., Chang, J., Coen, M., Cope, D., et al. (2016). Lab-on-a-drone: Toward pinpoint deployment of smartphone-enabled nucleic acid-based diagnostics for mobile health care. *Analytical Chemistry, 88*(9), 4651–4660.

Pupic, T. 2017. Innovation for impact: MENA startups are taking on the refugee crisis. *Entrepreneur.* Accessed 29 July 2018. https://www.entrepreneur.com/article/295806

Staruch, R., Beverly, A., Sarfo-Annin, J. K., & Rowbotham, S. (2018). Calling for the next WHO Global Health initiative: The use of disruptive innovation to meet the health care needs of displaced populations. *Journal of Global Health, 8*(1), 010303.

VSee Team in Iraq Kurdistan Refugee Telemedicine Clinic. 2014. *VSee.* Accessed 27 Aug 2018. https://vsee.com/blog/syrian-refugees-vsee-telemedicine-duhok/

Walji, M. (2015). Bringing telehealth to humanitarian settings. *CMAJ, 187,* E123–E124.

Zajc, T. Announcing the first deployment of the Iryo system: Improving healthcare for refugees. 2018. Accessed 26 Aug 2018. https://medium.com/iryo-network/announcing-the-first-deployment-of-the-iryo-system-improving-healthcare-for-refugees-bee8c441e7e6

Zarei, M. (2017). Portable biosensing devices for point-of-care diagnostics: Recent developments and applications. *TrAC Trends in Analytical Chemistry, 91,* 26–41.

Chapter 5
Humanitarian Innovation in the Modern Era: Ending Human Suffering

The ultimate goal humanitarian operations and innovation in general is to improve and enhance the human condition. With regard to humanitarian medicine and innovation, this is taken a step further with the fundamental element of alleviating human suffering and saving lives in the field. While the intentions are indeed very noble, innovation must be fostered by an evidence-based approach in conjunction with the tenets of human ingenuity. While indeed ideas open the door to novel interventions, if we are to make a true impact on the more than 63 million displaced refugees and conflict victims around the world, we must turn thought into action. The foundations provided in this book seek to take the ambiguity out of "innovation" and define its process, philosophy, and practical applications both now and in the future. Humanitarian innovation is perhaps the most extreme application of innovation, as it embraces the notion of creating something out of very limited resources. While indeed many may think that having limited resources serves as a detriment to the innovation process, we think quite the opposite. Resource-poor settings and unconventional environments employ people to think harder, deeper, and more efficiently. This creates a selection process whereby only the most practical and feasible ideas are embraced and pursued. At the end of the day, this breeds functional breakthroughs and innovations that display the same functional utility as those developed with ample resources. Furthermore, innovation is not embraced solely by highly educated professionals, researchers, academics, etc., but rather knows no bounds. The recipients and users of technological innovations often have as much or even more potential to innovate than the purveyors/creators of the innovation themselves. Ultimately, working together across geographic, socioeconomic, and geopolitical spectrums fosters dynamic creativity, enhanced problem-solving, as well as creates a better innovation. Human suffering is often viewed in the short-term in humanitarian operations; thus the interventions promoted are not suitable for perpetual application. By creating an innovation ecosystem that embraces the integration of an array of innovations ranging from 3D printing prosthetics to mHealth and

K. W. Ramadurai, S. K. Bhatia, *Reimagining Innovation in Humanitarian Medicine*, SpringerBriefs in Bioengineering, https://doi.org/10.1007/978-3-030-03285-2_5

blockchain technologies, we create infrastructure and human capital that will last into the future. This is where the notion of empowering refugee and conflict victim innovation comes into play. Who better to know the true realm of human suffering and the key to alleviating it than the actual people experiencing it firsthand. The future of humanitarian innovation lies not solely in the creation of novel technologies, devices, and interventions but more importantly in the knowledge transfer created and fostered. Health is fundamental element of the human condition, in which the bilateral transfer of knowledge is critical to empowering displaced individuals and promoting social equity.

5.1   Reworking Knowledge Transfer in the Humanitarian Ecosystem: Empowering Conflict Victim and Refugee Innovation

So far in this book, we have taken the term "innovation" and broken it down to the fundamental elements and processes that it encompasses specifically in relation to humanitarian medicine. Furthermore, we took it step further and created a functional taxonomy of the specific subcategorizations of innovation and its relevant contextual applications. But when it comes to conflict victim and refugee innovation, what type of innovation are we talking about? When it comes to fully empowering displaced populations, the need for bottom-up innovation becomes apparent. This represents a complete 360 from the standard humanitarian innovation process which is often top-down in nature. Bottom-up innovation represents the way in which crisis-affected communities engage in creative and dynamic problem-solving in order to adapt products, services, and processes in order to address challenges and create opportunities (Betts et al. 2015; Welling et al. 2010). Refugee populations are an excellent catalyst for bottom-up innovation, as they seek to improve and enhance the state of their condition. Given that the world now has more displaced people than at any time since World War II, fostering refugee innovation will be a key element in not only enhancing humanitarian medicine but the system as a whole. Bottom-up innovation by crisis-affected communities has become extremely under-recognized and utilized in any context. This is despite constant efforts to engage the capacities of these communities, but the reality is that a significant proportion of humanitarian innovation remains focused on improving organizational and agency response (Betts et al. 2015; Welling et al. 2010). But hope is not lost, formal acknowledgement of crisis-afflicted communities as stakeholders in humanitarian innovation is the first step in fostering a paradigm shift in aid and relief services. But where do we even begin?

Figure 5.1 identifies the interconnected initiatives that refugees as well as aid agencies have taken in creating a flow of information throughout the Za'atari refugee camp in Jordan. The refugee-led innovations are highlighted with stars, but what is interesting is the interconnection with external and international services or

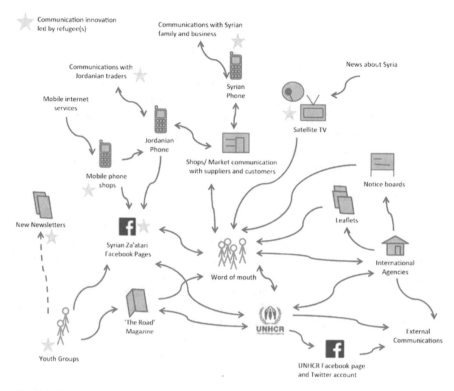

Fig. 5.1 Communications systems in Za'atari camp. (Betts et al. 2015)

projects. Within this communication schematic lies a powerful tool for refugee innovation—a network interface. This network serves as a powerful purveyor of information and knowledge dissemination in an almost instantaneous fashion. This informal network is the key to unlocking the true potential of refugee innovation and creating a function niche for innovation in the field. The key is tapping into this network's potential and utilizing it as a force for good. This informal network can easily be reconfigured to include aid agencies and nonprofits directly with projects in the field. This network can further serve as a platform for crowdsourced innovation within the refugee camp itself and rapidly accelerate problem-solution discovery.

The Za'atari refugee camp has been the host to a multitude of top-down efforts to introduce innovative products and processes by the international community. While these efforts have yielded mixed results, it is fascinating to note the informal infrastructure elements that have been created in the camp. The area the camp occupied was a barren desert and has now evolved into the fourth largest urban population in Jordan, housing more than 80,000 refugees (Betts et al. 2015). Basic services such as food, shelter, and utilities are provided by international agencies, but the camp itself is completely unconventional. The occupying Syrian refugees are known for their entrepreneurialism and have created their own functioning city-like area. One of the most prominent entities created by the refugees is the Shams-Élysées market, which has become a fixture in camp and extremely popular for the sale and distribution of goods and services (Fig. 5.2) (Betts et al. 2015). Specifically, the main market strip is host to more than 3000 Syrian-run stores, with the majority of shops selling food, as well as barber shops, beauty salons, phone repair shops, jewelry shops, and even foreign currency exchange (The Entrepreneurial Spirit of the 'Shams-Elysees'- Zaatari Refugee Camp 2018). The camp is host to an immense amount of human capital, which has spurred a range of innovations related to infrastructure development as well as the sale of goods and services. It has even led to social and gender equity opportunities whereby women in the camp have become entrepreneurs selling everything from food to hand crafted trinkets (Betts et al. 2015).

What we can see here is that refugees can create something out of literally nothing; thus their relative capacity to innovate truly knows no bounds. It is this example that makes refugee innovation a vital element in fostering innovation in humanitarian medicine. Harnessing the human capital of these individuals in conjunction with aid agency/organization support can lead to exciting developments. For example, oftentimes refugees are skilled in specific trades and also can include professionals such as doctors and engineers. Embracing the talents of these individuals can certainly create novel insights into the functional deployment of new frugally engineered technologies. Who better to know the problems related to healthcare access and delivery in refugee camps than the refugees themselves? This creates a full-circle dynamic of problem-solution discovery whereby refugees can functionally propose

Fig. 5.2 The Shams-Élysées market corridor in the Za'atari refugee camp. (The Entrepreneurial Spirit of the 'Shams-Elysees'- Zaatari Refugee Camp 2018)

a problem but also have the ability to create a feasible solution with the support of international or local agencies. The feedback loop created between refugees and the purveyors of new innovations such as mHealth blockchain or lab-on-a-drone technology is vital to enhancing the potential of these innovations to improve the human condition. The wants and needs of refugee communities lie in the elements of survival, social exchange, culture, and infrastructure as shown in Fig. 5.3 (Guthrie 2017). In looking closer at the figure, we can see that securing a water source, food distribution, shelter, security, and health services are the most vital elements during the first phase of crisis response, which is the emergency phase. The critical nature of securing health services should not only be taken into consideration during the emergency phase of a response. The reality is that health services will always be critical in refugee camps, years and years after the initial emergency response phase. Thus, it is absolutely vital the refugee innovation be fostered and developed as that will secure the future health and wealth of these individuals long into the future.

When it comes to facilitating refugee innovation, the key is to not only talk the talk but walk the walk. There is clearly an immense amount of untapped human capital and ingenuity, but how do we not only functionally tap this resource, but sustain it in creating real interventions? The first step, which we mentioned, is to first realize the capacity of crisis-afflicted communities to engage in innovation. Second, we need to understand the opportunities and barriers to the bottom-up innovation process in these camps. Third, a supportive and collaborative environment for innovation must be fostered by crisis-affected communities. Fourth, in order to propel the bottom-up innovation process, collaborative processes must be facilitated among multilateral agencies and nonprofit entities functioning in the refugee ecosystem, in order to fully support refugee innovation. Finally, while indeed innovation can be frugal in nature, the reality is that some sort of input capital is needed in order

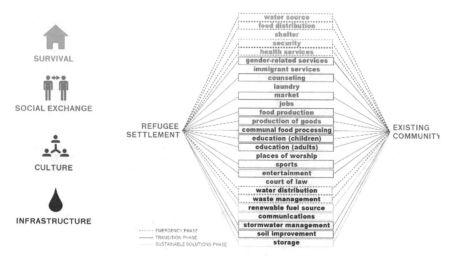

Fig. 5.3 The critical elements in emergency, transition, and sustainable solution phase of humanitarian interventions. (Guthrie 2017)

to garner momentum. Thus, the development of a sustainable humanitarian funding mechanism must be created in order support bottom-up innovation in the short and long terms. These steps do provide a relative path to fostering innovation in crisis-afflicted communities, but what are the precise enabling and constraining elements to propelling refugee innovation in these communities? Table 5.1 contains an excellent breakdown of these specific elements.

Upon reflection of Table 5.1, we can see that the tenets of refugee innovation must be fostered on an institutional, community, and individual level. Each one of these levels provides distinct enabling and constraining elements that must be overcome and embraced in order to break the barriers facing refugee innovation. With regard to humanitarian innovation, this becomes tricky, as many times medical allocation in crisis situations is top-down in nature. We advocate that this process can certainly be fostered in a top-down context during the initial emergency phase of humanitarian response, with the ultimate goal of transitioning to a sustainable bottom-up framework. The reality is that many humanitarian interventions are not in the scale of days or months, but rather years and even decades. As each year goes by, the interventional capacity of aid and relief agencies begins to slow due to fiscal and operational constraints, thus this provides an opportunity to develop refugee innovation. This innovation process serves to promote independence, personal and human capital development, as well as enhanced health outcomes. But where does refugee innovation fit in the conventional humanitarian network interface as shown in Fig. 5.4? When we look closer at Fig. 5.4, we can see that there are a myriad of agents that play a role in the humanitarian information network. Each one of these agents, i.e., NGOs, the public, academia, military, local government, private organizations, etc., has the distinct capacity for knowledge transfer, but the true challenges lies in the ability for these agents to facilitate an open transfer of knowledge and

Table 5.1 The enabling and constraint elements related to refugee innovation. Betts et al. (2015)

	Enablers	Constraints
Individual	Personal drive and motivation Existing skills or motivational to seek new skills Willingness to take risk Access to financial capital Local language skills Social networks	Loss of assets Psychosocial trauma Precarious and temporary legal status
Community	Local market access Communal support from others Community-led initiatives Encouragement and guidance	Local or national insecurity Xenophobia and discrimination
Institutional	Aid from international and local agencies Physical security provided by the state Right to work Right to register community-based organizations Access to public and private services	Lack of access to finance and banking Lack of full documentation Discrimination from authorities Lack of right to work Deportation

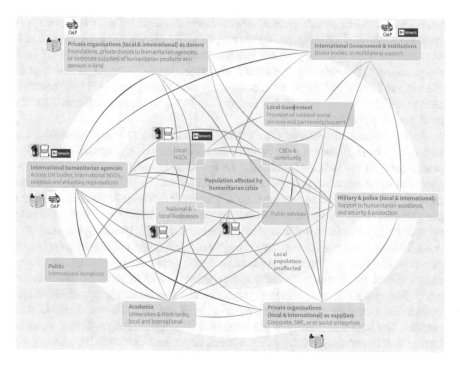

Fig. 5.4 The humanitarian network interface. (Betts and Bloom 2014)

information. When there are too many parts in a machine, the machines relative ability to break down becomes increased, and this analogy holds true for humanitarian response. But there is a defining element looked right in the middle and center of Fig. 5.4—this being the population affected by humanitarian crisis. This simple placement of crisis-afflicted communities in the middle of the network interface is an extremely important condition. Why? This is often where we see unilateral, rather than bilateral, transfer and exchange of knowledge and information among external agents when interacting with crisis-afflicted communities. This is an absolutely vital phenomenon that must be corrected in enhancing knowledge transfer and the ability to craft high-impact interventions. This interaction is also the functional basis for developing refugee innovation initiatives, right at the core of the network. This means that there is indeed an immense potential for refugee innovation processes to be integrated into this network, but getting external agents to formally recognize this proves to be challenging. The innovation process is integrative in nature and involves the functional input of stakeholders to identify, collaborate, and solve unconventional problems in the field. The refugee innovation process revolves around a central stakeholder—this being crisis-afflicted communities. The input from this centralized stakeholder can greatly influence the trajectory of running a lean response operation and can also shape the way patient and community medical needs are met. Furthermore, in this book, we have identified powerful innovations and agents of change that have yet to be formally incorporated into humanitarian medicine and greater network interface. This includes agents such as crowdsourcing and Wikicapital, which can radically disrupt the conventional network and provide a direct problem-solution discovery platform for crisis-afflicted communities to benefit from. It is open-knowledge transfer networks such as crowdsourcing and Wikicapital that add a completely new dimension to not only humanitarian medicine and innovation but the way we as human beings solve the world's most pressing problems.

As we mentioned previously, protecting human health is a key element of the human condition. Embracing the input of the patients receiving care with that of the practitioners in the field can open the door for novel frugally engineered technologies such as enhanced diagnostic tests, mHealth/eHealth platforms, telehealth, and an array of other promising technologies. The feedback loop between patient and practitioner provides a dynamic interface for problem-solution discovery, which can serve as an impetus for novel humanitarian medical devices, technologies, and solutions. Utilizing refugee innovation in conjunction with external agencies can serve as a powerful tool to solving unconventional problems in unconventional environments. In fact, this has recently been enacted in practice in Jordan's Za'atari refugee camp. The camp manager, Hovig Etyemezian, has independently spearheaded initiatives that seek to harness refugee innovation in all shapes and forms via the United Nations. The immense success of the more than 3000 Syrian-run small businesses is largely due to refugees driving market-based needs and the UNHCR offering the space to let these micro-entrepreneurs setup shop (How Jordan's Refugee Camps Became Hubs of Sci-Fi Tech and Booming Business 2016). This philosophy values human dignity, as well as the entrepreneurial forces of refuges to create and adapt to elements in the camp. The micro-entrepreneur is the impetus for refugee innovation; thus combining the entrepreneurial spirit of crisis-afflicted communities with that of the resources and additional human capital by external agencies can create a powerful force for good in propelling real innovation in humanitarian medicine.

5.2   The Future of Humanitarian Medicine and Creative Problem-Solving

The big question that all researchers, such as ourselves, as well as multilateral response agencies have is "what does the future of humanitarian aid and medicine hold?". The answer to this question is often derived in hyper-ambiguity, but we seek to provide a more succinct answer. In order to chart a course for the future, we must reflect on the past, in which there are "seven deadly sins" of humanitarian medicine that must not be embraced if we are to propel progress in the future (Welling et al. 2010). Table 5.2 shows these seven deadly sins, and while they are rather broad in scope and definition, they provide relative guidelines that seek to refine the way innovation and interventions in humanitarian medicine are fostered. Perhaps the most critical sin that can serve as an impediment to innovation in humanitarian medicine is that of Sin #2. Failing to match technology to local needs and abilities

Table 5.2 The seven sins of humanitarian medicine. Welling et al. (2010)

Sin #1: Leaving a mess behind
Sin #2: Failing to match technology to local needs and abilities
Sin #3: Failing of NGOs to cooperate and help each other and to cooperate and accept help from military organizations
Sin #4: Failing to have a follow-up plan
Sin #5: Allowing politics, training, or other distracting goals to trump service while representing the mission as "service"
Sin #6: Going where we are not wanted or needed and/or being poor guests
Sin #7: Doing the right thing for the wrong reason

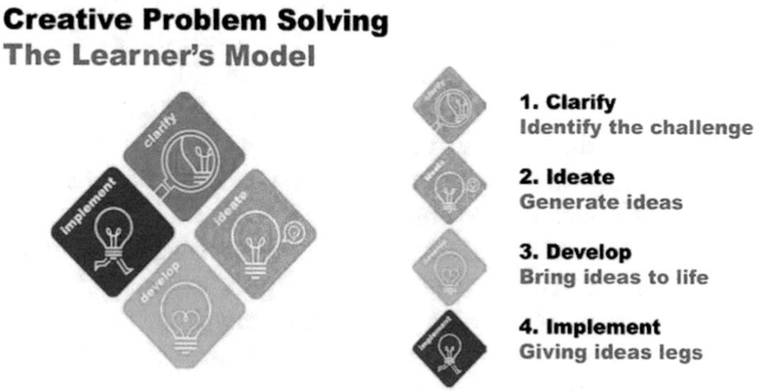

Creative Problem Solving
The Learner's Model

1. Clarify
Identify the challenge

2. Ideate
Generate ideas

3. Develop
Bring ideas to life

4. Implement
Giving ideas legs

Fig. 5.5 The four fundamental elements of the creative problem-solving process (Adapted from Puccio et al. 2011; Miller et al. 2011)

is the true "kryptonite" to any innovation being functional adapted and utilized to alleviate human suffering. If we create novel technologies and platforms, but they are not needed or wanted, we have lost the mission. This is why involving refugees and crisis-afflicted communities is so beyond vital in the humanitarian innovation process. Having the individuals that will be reaping the benefits of these innovations providing functional input makes sure that the innovation truly meets their needs. There is no bigger threat to the innovation paradigm than creating things that are not needed, functional, or applicable to their target environment and settings.

Upon realizing that Sin #2 has real implications for the future of humanitarian medicine, what do we do to stop the perpetuation of this problem? Well, we go back to the basics of creative problem-solving in the field. Figure 5.5 displays the four elements of the creative problem-solving process, which involves clarifying the challenge/problem, ideating ideas, developing and bringing these ideas to life, and implementing these ideas (Puccio et al. 2011; Miller et al. 2011). Each of the four steps of the CPS process involves both divergent and convergent thinking. Convergent thinking is a problem-solving technique that brings together different ideas from different stakeholders in order to determine the single best solution to a problem (Puccio et al. 2011; Miller et al. 2011). Divergent thinking is a problem-solving technique used to generate creative ideas by exploring many possible solutions that could be considered unusual, with the goal of aiming for quantity and the incubation of ideas (Puccio et al. 2011; Miller et al. 2011). This model is universally applicable to any agency, individual, or community and easily complements and integrates the innovation processes we explored earlier. What is even more critical in looking closer at the creative problem-solving paradigm is that different types of thinking garner different types of approaches and problem-solution discovery.

Specifically, Fig. 5.6 shows the concept of "design thinking" (DT), which is perhaps one of the most commonly deployed and effective thinking processes deployed in the creative problem-solving and innovation process (Loizou 2018). This further involves defining why the innovation is important, does it actually work, and how we can create it. This grounds the innovation and creative problem-solving process, so that feasible innovations can be derived with actual utility in the field. The main difference between the DT and the CPS processes is that design thinking is a far newer process, having been defined in 2003 (Loizou 2018).

The first stage involves "empathy," whereby consultations with experts and research are conducted and utilized as a springboard to address a problem/challenge. The second stage is "define," in which the individuals become aware of peoples' needs in accordance to the formula of user + need + insight (Loizou 2018). The third stage is "ideate," which involves the brainstorming of a plethora of ideas including the possible and impossible. The next stage is "prototype," which is a very rough and rapid portion of the design process. A prototype can be a sketch or model that can effectively convey an idea quickly. The last stage is "test," which is an iterative process that provides direct feedback in order to gauge what works and what does not in accordance with your user's needs. Figure 5.7 displays the factors that it

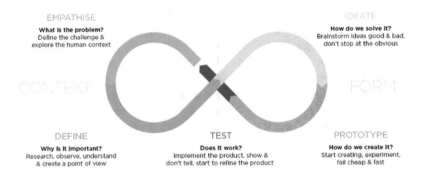

Fig. 5.6 Design thinking and deriving innovation. (Loizou 2018)

Fig. 5.7 The pieces of creating an innovation. (Basadur and Basadur 2018)

takes to create innovation results, which includes content, process, process skills, tools, and style (Basadur and Basadur 2018). But who will actually utilize these processes, skills, and problem-solving paradigms? These models serve as a reference point of reflection in the derivation of an innovation and make sure that it is grounded and directed toward helping it target user population. With regard to humanitarian medicine, this involves the formal acknowledge of a problem plaguing patients, engaging with the patients to understand their needs, then taking these needs, and creating a functional product via the design thinking process in order to alleviate human suffering.

When it comes to the future of creative problem-solving (CPS) in the humanitarian field, design thinking must be integrated in order to develop a holistic problem-solving and innovation process. This can be applied to the engineering of medical devices, mobile health platforms, and other health innovations that are deployed in the field. The future of frugal engineering and the development of low-cost medical devices and technologies lies in the integration of the CPS and DT processes in the creative product development and innovation pipeline. The combination of these creative thinking processes is known as the "creative process mash-up" (Fig. 5.8) (Puccio et al. 2018). This extremely novel creative thinking and problem-solving process was developed in 2018 and merges DT and CPS by blending the DT priority of a "user-based focus" with the CPS stage of "clarify" (Puccio et al. 2018). This novel process refines the creative thinking and problem-solving paradigm starting with the ability to "observe and define" in order to more deeply "understand" users (Puccio et al. 2018). The "develop" stage of CPS is changed to "experiment," as ideas are now actively and creatively "developed and validated" through iterative prototyping and openness to feedback from users (Puccio et al. 2018). Blending DT with the feedback of the CPS process further eliminates the uncertainties that are often present with the implementation of creative ideas and innovations, as well as the functional embrace of creative innovation and change in the field.

Oftentimes in humanitarian operations, the right partners are not brought together to achieve a common goal for the humanitarian sector (Delarue 2015). Furthermore, the needs of crisis-afflicted communities and populations were not analyzed properly, in which these individuals were not at the forefront of the process of defining the solution. The humanitarian is characterized as highly volatile and unpredictable, which further hinders external and private sector investing, making it difficult to facilitate the resources needed for the creation of innovation hubs (Delarue 2015).

Fig. 5.8 Creative process mash-up. (Adapted from Puccio et al. 2018)

When it comes to innovation in general, humanitarian agencies often cite lack of resources to justify their inability to come up with innovations that can scale (Delarue 2015). This brings the question, "how can humanitarian agencies save lives in an efficient and effective manner, if they do not invest in the development and scale of innovations?". This dilemma often favors inertia and prevents the development of new solutions, but nonetheless, we can break this paradigm via the embracement of a few key elements. The first and perhaps most critical is fostering an open problem-solving and innovation process that involves stakeholders of all backgrounds. Having platforms such as crowdsourcing can exponentially increase the efficacy and development of novel innovations. Furthermore, having this process start with the precise definition of a challenge, creation of solutions through rapid prototyping, user testing, iteration, and final prototyping for production is imperative (Delarue 2015). The next element involves the power of human-centered design (HCD), which is what we have discussed thoroughly in this work (Delarue 2015). This concept involves putting the end-user (in this case, crisis-afflicted communities) at the center of the solution definition. The closer these individual's needs are analyzed, the more successful the adoption of a solution/innovation is. The next element is the power of strategic partnerships, which is a prominent element in private sector collaborations among companies and academic institutes. This true nature of this concept in the humanitarian sector is not well embraced, yet without it there would be not novel solutions or innovations developed. The final element is the formal realization of venture philanthropy. Philanthropy provides the fiscal resources for required for rapid prototyping, lab testing, user testing, and iteration (Delarue 2015).

What we have seen throughout this book is that humanitarian medicine and the innovation processes that define the future development of novel medical devices and technologies are all innate to the human condition. With the increasing burden of chronic illness and disease in conjunction with increased conflict and disaster-afflicted displaced people, the time has come for innovation in humanitarian medicine. The fact of the matter is that while the intentions behind innovation may prove to be fruitful, the functional intervention and scaling of them present a true challenge for humanity. We all have the capacity to innovate, but it is what we do with our capacity to innovate that truly defines us.

References

Basadur, B., & Basadur, M. The process of innovation: Learning and inventing. Insights & Research. (2018). Accessed 02 Sept 2018. http://www.basadur.com/insightsresearch/OurThoughtsonCreativityandInnovation/TheProcessofInnovationLearningandInventing/tabid/168/Default.aspx

Betts, A., & Bloom, L. (2014). *Humanitarian innovation: The state of the art.* New York: United Nations Office for the Coordination of Humanitarian Affairs (OCHA).

Betts, A., Bloom, L., & Weaver, N. (2015). *Refugee innovation: Humanitarian innovation that starts with communities, Humanitarian Innovation Project.* Oxford: University of Oxford.

Delarue, O. (2015). 7 Powers for solving wicked humanitarian problems – UNHCR Innovation. *UNHCR Innovation.* Accessed 29 Aug 2018. http://www.unhcr.org/innovation/7-powers-for-solving-wicked-humanitarian-problems/

Guthrie, Sydney. 2017. Refugee camps of the future – Vignette Interactive. *Vignetteinteractive.com.* Accessed 30 Aug 2018. https://www.vignetteinteractive.com/refugee-camps-future/

How Jordan's refugee camps became hubs of sci-fi tech and booming business. 2016. *World Economic Forum* Accessed 3 Sept 2018. https://www.weforum.org/agenda/2016/09/jordan-refugee-camps-tech-business-united-nations/

Loizou, B. Design thinking. (2018). *The Visionary Owl.* Accessed 1 Sept 2018. http://www.billyloizou.com/thoughts/a-framework-for-innovation-designthinking

Miller, B., Vehar, J., Firestien, R., Thurber, S., & Nielsen, D. (2011). *Creativity unbound. An introduction to creative process* (5th ed.). Chicago: FourSight, LLC.

Puccio, G. J., Mance, M., & Murdock, M. C. (2011). *Creative leadership: Skills that drive change* (2nd ed.). London: Sage Publications.

Puccio, G. J., Cabra, J. F., & Schwagler, N. (2018). *Organizational creativity.* Buffalo: ICSC Press.

The entrepreneurial spirit of the 'Shams-Elysees'– Zaatari Refugee Camp. 2018. *1 Journey Festival.* Accessed 29 Aug 2018. https://www.onejourneyfestival.com/blog/2018/3/29/the-entrepreneurial-spirit-of-the-shams-elysees-zaatari-refugee-camp

Welling, D. R., Ryan, J. M., Burris, D. G., & Rich, N. M. (2010). Seven sins of humanitarian medicine. *World Journal of Surgery, 34*(3), 466–470.

Printed in the United States
By Bookmasters